Transplantation Drug Manual

Fifth Edition

T0346446

Transplantation Drug Manual

Fifth Edition

John Pirsch, MD
Professor of Medicine and Surgery
Director of Medical Transplantation
University of Wisconsin
Madison, Wisconsin, USA

Hans Sollinger, MD, PhD
Folkert O. Belzer Professor of Surgery
Chairman, Division of Organ Transplantation
University of Wisconsin
Madison, Wisconsin, USA

Xiaolan Liang, PharmD, BCPS
Clinical Pharmacist
School of Pharmacy Instructor
Transplantation Specialist
University of Wisconsin
Madison, Wisconsin, USA

CRC Press
Taylor & Francis Group
Boca Raton London New York

CRC Press is an imprint of the
Taylor & Francis Group, an **informa** business

TRANSPLANTATION DRUG MANUAL, 5TH EDITION

First published 2007 by Landes Bioscience

Published 2018 by CRC Press
Taylor & Francis Group
6000 Broken Sound Parkway NW, Suite 300
Boca Raton, FL 33487-2742

© 2007 by Taylor & Francis Group, LLC
CRC Press is an imprint of Taylor & Francis Group, an Informa business

No claim to original U.S. Government works

ISBN 13: 978-1-57059-698-8 (pbk)

Visit the Taylor & Francis Web site at
http://www.taylorandfrancis.com

and the CRC Press Web site at
http://www.crcpress.com

While the authors, editors, and publisher believe that drug selection and dosage and the
specifications and usage of equipment and devices, as set forth in this book, are in accord
with current recommendations and practice at the time of publication, they make no
warranty, expressed or implied, with respect to material described in this book. In view of
the ongoing research, development, changes in governmental regulations and the rapid
accumulation of information relating to the biomedical sciences, the reader is urged to
carefully review and evaluate the information provided herein.

Library of Congress Cataloging-in-Publication Data

Pirsch, John.
  Transplantation drug manual / John Pirsch, Hans Sollinger, Xiaolan Liang. -- 5th ed.
    p. ; cm.
  Includes bibliographical references.
  ISBN-13: 978-1-57059-698-8
  1. Drugs--Handbooks, manuals, etc. 2. Transplantation of organs, tissues, etc.--Handbooks,
manuals, etc. 3. Transplantation of organs, tissues, etc.--Complications--Handbooks, manuals,
etc. I. Sollinger, Hans. II. Liang, Xiaolan. III. Title.
  [DNLM: 1. Immunosuppressive Agents--pharmacology--Handbooks. 2. Transplantation--
Handbooks. 3. Anti-Infective Agents--pharmacology--Handbooks. 4. Postoperative Complica-
tions--drug therapy--Handbooks. WO 39 P672t 2007]
  RM301.12.P57 2007
  615.5'8--dc22
                                                    2007024909

Table of Contents

Introduction

The field of transplantation continues to evolve. In the years since the first publication of the *Transplantation Drug Pocket Reference Guide*, the therapeutic armamentarium for transplantation has grown. The introduction of new agents continues to enhance our ability to improve the short- and long-term success of transplantation.

The fifth edition of the guide, now entitled *Transplantation Drug Manual*, includes information on agents approved for use in transplant recipients. As in previous editions, we compiled practical information on the wide array of pharmaceutical agents currently available—both those used for immunosuppression and those used to minimize posttransplant complications. The agents described here are the most frequently prescribed drugs in the Transplant Service at the University of Wisconsin. We hope you find this information to be useful and practical in managing the transplant patient.

Chapter 1
Working Guide to Immunosuppression

Current Agents Overview

- Antithymocyte Globulin (equine)—Atgam
- Antithymocyte Globulin (rabbit)—Thymoglobulin
- Muromonab-CD3
- Basiliximab
- Daclizumab
- Alemtuzumab
- Rituximab
- Immunoglobulin
- Azathioprine
- Cyclophosphamide
- Cyclosporine-A (non-modified)
- Cyclosporine Capsules and Oral Solution (modified)
- Methylprednisolone
- Mycophenolate Mofetil
- Mycophenolate Sodium Delayed-Release
- Prednisone
- Tacrolimus
- Sirolimus

Current Agents Overview

Immuno-suppression

Agent	Dosage
Antithymocyte Globulin (Atgam)	*Delaying Onset of Allograft Rejection* • Fixed dose of 15 mg/kg for 14 days, then every other day for 14 days for a total of 21 doses in 28 days • First dose should be administered within 24 hours before or after transplantation *Treatment of Rejection* • 10-15 mg/kg/d IV for 8-14 days, then every other day up to 21 doses
Antithymocyte Globulin (rabbit)	*Treatment of Rejection* • 1.5 mg/kg/d for 7 to 14 days IV
Muromonab-CD3	• 5 mg/d IV for 10 to 14 days
Basiliximab	• 20 mg within 2 hours of transplantation surgery and repeated 4 days after transplantation
Daclizumab	• 1 mg/kg/dose for 5 doses. The first dose within 24 hours of transplantation, then at intervals of 14 days for 4 doses
Alemtuzumab	• 30 mg intravenously on the day of transplantation surgery and another 30 mg may be given on post-operation day 1
Rituximab	• 375 mg/m² (with a maximum dose of 750 mg) intravenously, may repeat one more time 2 weeks later
Immunoglobulin	• 100 mg/kg (maximum 7000 mg) IV • 2 gram/kg (maximum of 140 grams) IV
Azathioprine	• 3 mg/kg/d to 5 mg/kg/d single dose given at time of transplantation • 1 mg/kg/d to 3 mg/kg/d for maintenance • Dose usually adjusted depending on WBCs • Lower doses should be considered in presence of renal dysfunction
Cyclophosphamide	• 2 mg/kg/d to 3 mg/kg/d is initial recommended dose, but is rapidly reduced due to toxicity
Cyclosporine-A (non-modified)	*Gelatin Capsules & Oral Solution* • 15 mg/kg single dose given 4 to 12 hours prior to transplantation • Usual starting dose in 4-10 mg/kg in bid dosing. Dose is titrated to achieve whole blood levels of 200-300 mg/mL *IV Infusion* • 5 mg/kg to 6 mg/kg single dose given 4 to 12 hours prior to transplantation • Single daily IV dose continued postoperatively until patient can tolerate oral formulations
Cyclosporine Capsules and Oral Solution (modified)	**NOTE: Neoral, Gengral, and Sandimmune are not bioequivalent and should not be used interchangeably. Neoral has increased bioavailability, and this should be taken into consideration when making dosing decisions.** • Daily dose should be given as two divided doses on a consistent schedule

Continued ...

Antimicrobials

Cardiovascular Agents

Antiosteo-porosis Agents

Antiplatelets

Diabetes Agents

Ulcer Treatment

Diuretics

Other Agents

Current Agents Overview

Agent	Dosage

Cyclosporine Capsules and Oral Solution (modified) (continued)	*Newly Transplanted Patients*
	• Initial dose of Neoral should be the same as a Sandimmune dose. Suggested initial doses include:
	  – 9 ± 3 mg/kg/day for kidney transplant patients
	  – 8 ± 4 mg/kg/day for liver transplant patients
	  – 7 ± 3 mg/kg/day for heart transplant patients
	• The dose is then subsequently adjusted to achieve a predefined cyclosporine blood concentration
	
	Conversion from Sandimmune to Neoral
	• Neoral should be started with the same daily dose as was previously used with Sandimmune (1:1 dose conversion)
	• The Neoral dose should then be adjusted to achieve preconversion cyclosporine blood trough concentrations
	• Until cyclosporine blood trough concentrations reach preconversion levels, monitoring should be undertaken every 4 to 7 days

Methylprednisolone	*Induction* • 250 mg to 1000 mg at time of transplantation and for next 2 to 3 doses *Taper* • Start at 2 mg/kg/d, taper to a range of 0.15 mg/kg/d to 0.2 mg/kg/d after one year *Attenuation of Cytokine Release Syndrome* • 8 mg/kg given 1 to 4 hours prior to first injection of muromonab-CD3
Mycophenolate Mofetil	Initial dose should be given within 72 hours following transplantation • 1 gram twice a day used in combination with corticosteroids and cyclosporine
Mycophenolate Sodium	• 720 mg twice daily used in combination with other immuno-suppressants
Prednisone	*Maintenance—Adults* • 0.1 mg/kg/d to 2 mg/kg/d usually given once daily *Maintenance—Pediatric* • 0.25 mg/kg/d to 2 mg/kg/d or 25 mg/m^2 to 60 mg/m^2 usually given daily or on alternate days
Tacrolimus	*IV Infusion* • 0.03 to 0.05 mg/kg/d as a continuous infusion *Capsules* • 0.15 mg/kg/d to 0.30 mg/kg/d administered in 2 divided daily doses every 12 hours • First dose should be given 8 to 12 hours after discontinuing IV infusion
Sirolimus	*Adults and pediatric patients (>12 years old and >40 kg)* • Loading dose 6 mg orally followed by 2 mg daily maintenance dose *Pediatric (>12 years and <40 kg)* • Loading dose: 3 mg/m^2 orally followed by 1 mg/m^2 daily maintenance dose

Antithymocyte Globulin (equine)

Immuno-suppression

Brand Name	Atgam®
Company	Pharmacia & Upjohn, Inc.
Class	• Immunosuppressant gamma globulin, primarily monomeric IgG, from hyperimmune serum of horses immunized with human thymic lymphocytes
Mechanism of Action	• Antibodies of multiple specificities interact with lymphocyte surface antigens, depleting numbers of circulating, thymus-dependent lymphocytes and interfering with cell-mediated and humoral immune responses
Indications	• Management of renal allograft rejection • Adjunct to other immunosuppressive therapy to delay the onset of the first rejection episode
Contraindication	• Hypersensitivity to Atgam or any other equine gamma globulin preparation
Warnings	• Should be administered in facilities equipped and staffed with adequate laboratory and supportive medical resources • Immunosuppressive activity may vary from lot to lot • Potential for the transmission of infectious agents • Treatment should be discontinued if the following occur: • Symptoms of anaphylaxis • Thrombocytopenia • Leukopenia
Special Precautions	• Risk of infection, leukopenia, and thrombocytopenia • Safety and effectiveness demonstrated only in patients who received concomitant immunosuppression • Pregnancy Category C
Adverse Reactions	• Fever (1 patient in 3) • Chills (1 patient in 7) • Leukopenia (1 patient in 7) • Dermatologic reactions (1 patient in 8) • Thrombocytopenia (1 patient in 9) *Reported in >1%, but <5% of Patients* • Arthralgia • Chest and/or back pain • Clotted A/V fistula • Nausea and/or vomiting • Night sweats • Pain at infusion site • Peripheral thrombophlebitis • Stomatitis
Drug Interaction	• Dextrose Injection, USP
Formulation	• 5 mL ampule containing 50 mg/mL
Dosage	*Delaying Onset of Allograft Rejection* • Fixed dose of 15 mg/kg for 14 days, then every other day for 14 days for a total of 21 doses in 28 days • First dose should be administered within 24 hours before or after transplantation *Treatment of Rejection* • 10 mg/kg/d IV for 8-14 days, then every other day up to 21 doses Dose should be infused at least over 4 hours, through a 0.2-1 micron filter

Editors' Notes:

    Atgam® and Thymoglobulin® are the only two polyclonal antilymphocyte preparations which are currently available. Atgam® is an immunoglobule against lymphocytes prepared in horses; Thymoglobulin® is prepared in rabbits. Atgam® causes less leukopenia than Thymoglobulin®. Both agents have been used to prevent acute rejection after transplantation and to treat acute rejection episodes.

    In the latest SRTR report, antithymocyte globulin (rabbit) was the most common agent used for induction therapy (37% of patients receiving induction). Meier-Kriesche et al. AJT 2006; 6(Part 2):1111-1131.

Antimicrobials · Cardiovascular Agents · Antiosteoporosis Agents · Antiplatelets · Diabetes Agents · Ulcer Treatment · Diuretics · Other Agents

Antithymocyte Globulin (rabbit)

Brand Name	Thymoglobulin®
Company	Genzyme
Class	• Immunosuppressant gamma globulin, obtained by immunization of rabbits with human thymocytes
Mechanism of Action	• Antibodies of multiple specificities interact with lymphocyte surface antigens, depleting numbers of circulating T lymphocytes and modulating T-lymphocyte activation, homing and cytotoxic processes
Indication	• Treatment of acute renal allograft rejection in combination with other immunosuppressants
Contraindication	• History of anaphylaxis or allergy to Thymoglobulin or rabbit proteins, or acute viral illness
Warnings	• Should be administered in facilities equipped and staffed with adequate laboratory and supportive medical resources • Anaphylaxis has been reported • Thrombocytopenia or neutropenia may result but are reversible with dose reduction or discontinuance
Special Precautions	• Risk of infections, leukopenia, thrombocytopenia, lymphoma, post-transplant lymphoproliferative disease or other malignancies • Pregnancy Category C
Adverse Reactions	• Fever • Chills • Leukopenia • Pain/ abdominal pain • Headache • Thrombocytopenia • Dyspnea • Malaise • Dermatologic reactions
Drug Interaction	• None reported
Formulation	• 25 mg vial of lyophilized powder
Dosage	• 1.5 mg/kg/d for 7 to 14 days IV, diluted and infused through a 0.22 micron filter into a high flow vein • First dose infused over a minimum of 6 hours, subsequent doses over a minimum of 4 hours

Editors' Notes:

Antithymocyte globulin (rabbit) is approved for the reversal of acute rejection. A double-blind, randomized trial of Thymoglobulin vs Atgam was conducted in 163 renal recipients with rejection. Thymoglobulin had a higher reversal rate than Atgam (88% vs 76%, p=0.027). Transplantation 66:29-37, July 15, 1998.

A recent study of high-risk renal transplants compared Thymoglobulin® induction with Simulect. The overall risk of rejection, delayed graft function and graft loss was statistically less frequent with Thymoglobulin®. (Brennan DC et al. Rabbit antithymocyte globulin versus basiliximab in renal transplantation. N Engl J Med 2006; 355(19):1967-1977.

Immuno-suppression · Antimicrobials · Cardiovascular Agents · Antiosteoporosis Agents · Antiplatelets · Diabetes Agents · Ulcer Treatment · Diuretics · Other Agents

Muromonab-CD3

Immuno-suppression

Brand Name	**ORTHOCLONE OKT®3**
Company	Ortho Biotech Inc.
Class	• Immunosuppressive monoclonal antibody with singular specificity to CD3 antigen of human T cells
Mechanism of Action	• Blocks the function of CD3 molecule in the membrane of human T cells, which has been associated in vitro with the antigen recognition structure of human T cells that is essential for signal transduction
Indications	• Treatment of acute renal allograft rejection as soon as it is diagnosed • Treatment of steroid-resistant acute cardiac allograft rejection • Treatment of steroid-resistant acute hepatic allograft rejection
Contraindications	• Hypersensitivity to muromonab-CD3 and/or any product of murine origin • Antimurine antibody titers ≥1:1000 • Uncompensated heart failure or fluid overload • History of seizures • Determined to be or suspected of being pregnant or breast-feeding
Warnings	• Cytokine release syndrome, ranging from a mild, self-limited, flu-like syndrome to a less-frequently reported severe, life-threatening shock-like reaction, has been associated with first few doses • Can be attenuated by premedicating with methylprednisolone, 8 mg/kg, given 1 to 4 hours prior to first injection • Anaphylactic reactions • Neuropsychiatric events, including seizures, encephalopathy, cerebral edema, aseptic meningitis, and headache • Risk of infection and neoplasia • Patients should be managed in a facility equipped and staffed for cardiopulmonary resuscitation
Special Precautions	• Clear chest X-ray and weight restriction of ≤3% above patient's minimum weight during the week prior to injection • If patient's temperature exceeds 37.8°C, it should be lowered with antipyretics before each injection • Periodic monitoring to ensure muromonab-CD3 levels >800 ng/mL and CD3+ cell levels <25 cell/mm³ • Potentially serious signs and symptoms with immediate onset are likely due to hypersensitivity and therapy should be discontinued • Antiseizure precautions should be undertaken • Patient should be monitored for signs of infection and/or lympho-proliferative disorders • Antimurine antibody titers should be monitored after therapy with muromonab-CD3 • As with other immunosuppressives, arterial or venous thromboses of allografts and other vascular beds have been reported • Pregnancy Category C

Continued ...

Antimicrobials

Cardiovascular Agents

Antiosteo-porosis Agents

Antiplatelets

Diabetes Agents

Ulcer Treatment

Diuretics

Other Agents

Immuno-suppression

Antimicrobials

Cardiovascular Agents

Antiosteo-porosis Agents

Antiplatelets

Diabetes Agents

Ulcer Treatment

Diuretics

Other Agents

Muromonab-CD3

Adverse Reactions	• Pyrexia • Chills • Dyspnea • Nausea and vomiting • Chest pain • Diarrhea • Tremor • Wheezing • Headache • Hypersensitivity reactions	• Tachycardia • Rigor • Hypertension • Infection with herpes simplex virus, cytomegalovirus, *Staphylococcus epidermidis*, *Pneumocystis carinii*, *Legionella*, *Cryptococcus*, *Serratia*, and gram-negative bacteria • Posttransplant lympho-proliferative disorders
Drug Interactions	• Indomethacin alone and in conjunction with muromonab-CD3 has been associated with CNS effects • Corticosteroids alone and in conjunction with muromonab-CD3 have been associated with psychosis and infection • Azathioprine alone and in conjunction with muromonab-CD3 has been associated with infection and malignancies • Cyclosporine-A alone and in conjunction with muromonab-CD3 has been associated with seizures, encephalopathy, infection, malignancies, and thrombotic events	
Formulation	• 5 mL ampule containing 5 mg	
Dosage	• 5 mg/d IV for 10 to 14 days	

Editors' Notes:

Muromonab-CD3 is used infrequently to treat or prevent rejection in the present era. It has significant toxicity due to the cytokine defense syndrome (CDS).

Evidence suggests that giving 250 mg of methylprednisolone 6 hours prior and 1 hour prior to administration of muromonab-CD3 can decrease the incidence of cytokine release syndrome (CRS). Pentoxifylline has not been shown to attenuate CRS. Indomethacin, however, can be effective.

If CD3 levels remain high during therapy with muromonab-CD3, the dose may be increased or one may switch to an alternate antilymphocyte preparation.

Adjunctive prophylaxis with ganciclovir in CMV-positive patients or in recipients of CMV-positive organs may be effective in reducing risk of CMV infection.

The use of muromonab-CD3 in transplantation is minimal for induction or treatment of rejection.

Basiliximab

Brand Name	Simulect®
Company	Novartis
Class	• Immunosuppressive chimeric monoclonal antibody, specifically binds to and blocks the interleukin-2 receptor alpha chain on the surface of activated T- lymphocytes
Mechanism of Action	• Acts as an IL-2 receptor antagonist by binding with high affinity to the alpha chain of the IL-2 receptor complex and inhibits IL-2 binding • Competitively inhibits IL-2 mediated activation of lymphocytes
Indications	• Prophylaxis of acute renal allograft rejection when used as part of an immunosuppressive regimen that includes steroids and cyclosporine
Contraindications	• Hypersensitivity to basiliximab or any component of the formulation
Warnings	• Should be administered in facilities equipped and staffed with adequate laboratory and supportive medical resources • Administration of proteins may cause possible anaphylactoid reactions (none reported) • Immunosuppressive therapies increase risk for lymphoproliferative disorders and opportunistic infections (incidence in basiliximab-treated patients is similar to placebo)
Special Precautions	• Long-term effect and re-administration after initial course has not been studied • Pregnancy Category B
Adverse Reactions	• Similar to placebo-treated patients
Drug Interactions	• None reported
Formulation	• 10 mg vial of lyophilized powder • 20 mg vial of lyophilized powder
Dosage	• 20 mg within 2 hours of transplantation surgery and repeated 4 days after transplantation • Children 2 to 15 years is 12 mg/m², to a maximum of 20 mg/dose

Editors' Notes:

Basiliximab was approved after a study of 380 cadaver transplant recipients on cyclosporine and prednisone was completed. Basiliximab recipients had a rejection incidence of 29.8% compared to placebo (44%, p=0.01). Steroid-resistant rejection was also lower. Lancet 350(9086):1193-1998, October 25, 1997.

The clinical experience with Basiliximab had confirmed the results of this study with excellent efficacy and low toxicity.

A recent study of high-risk renal transplants compared Thymoglobulin® induction with Simulect. The overall risk of rejection, delayed graft function and graft loss was statistically less frequent with Thymoglobulin®. Brennan DC et al. Rabbit antithymocyte globulin versus basiliximab in renal transplantation. N Engl J Med 2006 Nov 9; 355(19):1967-1977.

Antimicrobials

Cardiovascular Agents

Antiosteo-porosis Agents

Antiplatelets

Diabetes Agents

Ulcer Treatment

Diuretics

Other Agents

Daclizumab

Brand Name	Zenapax®
Company	Roche Laboratories
Class	• Immunosuppressive humanized monoclonal antibody, specifically binds to and blocks the interleukin-2 receptor alpha chain on the surface of activated T- lymphocytes
Mechanism of Action	• Acts as an IL-2 receptor antagonist by binding with high affinity to the alpha chain of the IL-2 receptor complex and inhibits IL-2 binding • Competitively inhibits IL-2 mediated activation of lymphocytes
Indications	• Prophylaxis of acute renal allograft rejection when used as part of an immunosuppressive regimen that includes steroids and cyclosporine
Contraindication	• Hypersensitivity to daclizumab or any component of the formulation
Warnings	• Should be administered in facilities equipped and staffed with adequate laboratory and supportive medical resources • Administration of proteins may cause possible anaphylactoid reactions (none reported) • Immunosuppressive therapies increase risk for lymphoproliferative disorders and opportunistic infections (incidence in daclizumab-treated patients is similar to placebo)
Special Precautions	• Long-term effect and re-administration after initial course has not been studied • Pregnancy Category C
Adverse Reactions	• Similar to placebo-treated patients
Drug Interactions	• None reported
Formulation	• 5 mL vial containing 25 mg
Dosage	• 1 mg/kg/dose for 5 doses, the first dose within 24 hours of transplantation, then at intervals of 14 days for four doses. Dilute with 50 mL normal saline over 15 minutes.

Editors' Notes:

Daclizumab was approved after a study of 260 cadaver recipients on azathioprine, cyclosporine and prednisone was completed. Daclizumab recipients had a rejection incidence of 22% compared to placebo (35%, p<0.03). New Engl J Med 1998; 338(3):161-165.

Recent data from a kidney pancreas induction study suggests that 2 doses of Daclizumab (2 mg/kg) at day 0 and day 14 is equivalent to 5 doses of 1 mg/kg every 14 days. (Stratta AJ, Alloway RR, Hodge E et al. A multicenter, open-label, comparative trial of two Daclizumab dosing strategies vs. no antibody induction in combination with tacrolimus, mycophenolate mofetil, and steroids for the prevention of acute rejection in simultaneous kidney-pancreas transplant recipients: interim analysis. Clin Transplant ©2002; 16(1):60-8.)

Alemtuzumab

Immuno-suppression

Antimicrobials

Cardiovascular Agents

Antiosteo-porosis Agents

Antiplatelets

Diabetes Agents

Ulcer Treatment

Diuretics

Other Agents

Brand Name	Campath®
Company	Genzyme
Class	• Immunosuppressive recombinant DNA-derived humanized monoclonal antibody that is directed against the cell surface glycoprotein, CD52, on the surface of normal and malignant B and T lymphocytes, NK cells, monocytes, macrophages, and tissues of the male reproductive system
Mechanism of Action	• Targets and binds to CD52 antigen present on the surface of lymphocytes and induces cell lysis
Application in Transplantation	• Adjunct to other immunosuppressive therapy to delay onset of first rejection episode
Contraindications	• Active systemic infection • Hypersensitivity to alemtuzumab or components • Underlying immunodeficiency
Warnings	• Serious and in rare cases fatal hematological toxicities have been observed • Serious and sometimes fatal infusion related reactions may occur • Increased risk for serious/fatal infections
Special Precautions	• Hematological toxicities: single doses greater than 30 mg or cumulative doses greater than 90 mg/week should not be administered due to higher incidence of pancytopenia (weekly blood and platelet counts recommended) • Development of antibodies to alemtuzumab has occurred • Infusion related reactions can be managed by premedication with acetaminophen and diphenhydramine • Live attenuated vaccines should be avoided during alemtuzumab therapy • Anti-infective prophylaxis is recommended against herpes and Pneumocystis • Pregnancy Category C
Adverse Reactions	• Infusion related reactions:hypotension/fevers/rigor • Rash/urticaria • Nausea/vomiting/diarrhea • Pancytopenia/marrow dysplasia • Anemia • Neutropenia • Idiopathic thrombocytopenic purpura • Thrombocytopenia • Headache/dysesthesia/dizziness/tremor
Drug Interactions	• None reported
Formulations	• 1 mL vials containing 30 mg
Dosage	• 30 mg intravenously on the day of transplantation surgery and another 30 mg may be given on post-operation day 1

Editors' Notes:

Lymphocyte depletion is one of the proposed mechanisms to attain tolerance post-transplantation. Alemtuzumab causes profound lymphocyte depletion and is being used as an induction agent in kidney and kidney-pancreas. There are many unanswered questions about its use including a possible measured risk of antibody-mediated rejection and the long-term consequences of T-cell depletion leading to opportunistic infections and cancer.

Rituximab

Brand Name	Rituxan®
Company	Genentech
Class	• Immunosuppressive chimeric murine/human monoclonal antibody directed against the CD20 antigen found on the surface of B lymphocytes
Mechanism of Action	• Binds to the CD20 antigen on B lymphocytes and mediates B-cell lysis through complement-dependent cytotoxicity, antibody-dependent cell mediated cytotoxicity and apoptosis
Application in Transplantation	• Adjunct to other immunosuppressive therapy to treat antibody-mediated rejection • Adjunct to other immunosuppressive therapy in desensitization therapies for highly sensitized recipients and recipients of ABO incompatible transplants
Contraindications	• Known anaphylaxis or IgE-mediated hypersensitivity to murine proteins • Hypersensitivity to rituximab or any of its components
Warnings	• Fatal infusion reactions have been observed (approximately 80% of these reactions is associated with the first infusion) • Hepatitis B reactivation with fulminant hepatitis has been reported
Special Precautions	• Patients require close monitoring during infusion (reducing infusion rate or discontinuation of infusion may be needed) Regular monitoring of complete blood counts and platelet counts • Development of human anti-chimeric antibodies (<1%) • Safety with using live attenuated vaccines in patients receiving rituximab has not been studied and is not recommended • Patients of childbearing potential should use effective contraceptive methods during treatment and for up to 12 months following therapy • Pregnancy Category C
Adverse Reactions	• Infusion related reactions: hypotension (10%) /arrhythmia/fevers/rigor • Fatal mucocutaneous reactions • Rash/urticaria • Hyperglycemia (9%) • Peripheral edema (8%) • Nausea/vomiting/diarrhea • Pancytopenia/marrow dysplasia (rare) • Lymphopenia • Anemia • Leukopenia • Neutropenia • Thrombocytopenia • Reactivation of Hepatitis B infections • Headache/asthesia/dizziness
Drug Interactions	• None reported
Formulations	• 10 mg/mL in either 100 mg (10 mL) or 500 mg (50 mL) single-use vials
Dosage	• 375 mg/m^2 (with a maximum dose of 750 mg) intravenously, may repeat one more time 2 weeks later

Editors' Notes:

Rituximab has emerged as an adjunct theraphy for antibody-mediated rejection, desensitization protocols and in ABO-incompatible transplants. It is also highly effective as a first-time agent in the treatment of PTLD. Unfortunately, there are no randomized, controlled trials demonstrating efficacy. (Becker YT et al. The emerging role of rituximab in organ transplantation. Transpl Int 2006; 19(8):621-628.)

Immunoglobulin

Immuno-suppression

Antimicrobials

Cardiovascular Agents

Antiosteo-porosis Agents

Antiplatelets

Diabetes Agents

Ulcer Treatment

Diuretics

Other Agents

Brand Name/ Company	• Gammagard S/D®, Gammagard®, Polygam S/D®/*Baxter Healthcare Corporation* • Venoglobulin-S®, Venoglobulin-I/*Alpha Therapeutic Corporation* • Gamimun-N5%®, Gamimune-N 5% S/D®, Gaminume-N 10%®, Gaminume-N 10% S/D®/*Bayer Corporation* • Gammar-P-IV®/*Centeon, L.L.C* • Iveegam®/*Baxter Healthcare Corporation* • Sandoglobulin®/*Novartis Pharmaceutical Corporation*
Class	• Antibody products containing concentrated form of IgG antibodies that regularly occur in the donor population
Mechanism of Action	• Immunomodulatory effects possibly due to complement absorption, down regulation of immunoglobulin production, enhancement of suppression cells, or inhibition of lymphocyte proliferation and reduction of IL-1 production
Application in Transplantation	• Adjunct to other immunosuppressive therapy to treat antibody-mediated rejection episodes
Contraindications	• Hypersensitivity to human immunoglobulin • Selective IgA deficiency
Warnings	• Rate of infusion may cause a precipitous fall in blood pressure and a clinical picture of anaphylaxis even in the absence of hypersensitivity
Special Precautions	• Pregnancy Category C
Adverse Reactions	• Flushing • Urticaria • Slight elevation in blood pressure • Nausea/vomiting • Fever • Leg cramps • Lightheadedness • Headache • Chills • Backaches • Fatigue
Drug Interactions	• Drug interactions have not been evaluated. However, it is recommended that immunoglobulin be administered separately from other agents
Formulations	• 0.5 g single use bottle
	• 1 g vial • 2.5 g single use bottle • 3 g vial • 5 g single use bottle • 6 g vial • 10 g single use bottle • 10 mL/50 mL/100 mL/250 mL
Dosage	• 100 mg/kg (maximum 7000 mg) IV • 2 gram/kg (maximum of 140 grams) IV

Editors' Notes:

Immunoglobulin therapy is used as an immunomodulator for antibody-mediated rejection and desensitization protocols. The best review paper is: Jordan S et al. Utility of intravenous immune globulin in kidney transplantation: efficacy, safety, and cost implications. Am J Transplant 2003; 3(6):653-654.

Azathioprine

Brand Name	Imuran®
Company	Glaxo Wellcome Inc.
Class	• Immunosuppressive antimetabolite (6-mercaptopurine)
Mechanism of Action	• Interferes with DNA and RNA synthesis, thereby inhibiting differentiation and proliferation of both T and B lymphocytes
Indication	• Adjunct for the prevention of rejection in renal homotransplantation
Clinical Experience	• 35% to 55% 5-year patient survival in over 16,000 transplantations
Contraindication	• Hypersensitivity to azathioprine
Warnings	• Increased risk of neoplasia • Severe myelosuppression • Serious infection • Pregnancy Category D
Special Precautions	• Gastrointestinal hypersensitivity reaction • Periodic blood counts may be needed
Adverse Reactions	• Leukopenia (>50%) • Infection (20%) • Nausea and vomiting (NA) • Neoplasia (3.3%) • Hepatotoxicity (NA)
Drug Interactions	• Allopurinol • Agents affecting myelopoiesis • Angiotensin converting enzyme inhibitors
Formulations	• 50 mg scored tablets • 20 mL vial containing 100 mg azathioprine
Dosage	• 3 mg/kg/d to 5 mg/kg/d single dose given at time of transplantation • 1 mg/kg/d to 3 mg/kg/d for maintenance • Dose usually adjusted depending on WBCs • Lower doses should be considered in presence of renal dysfunction

Editors' Notes:

Combined administration of azathioprine and allopurinol may result in severe pancytopenia. The dose of azathioprine should be reduced by two-thirds when given with allopurinol. Azathioprine rarely causes liver dysfunction, frequently manifested by an isolated rise in ALT and bilirubin. With the introduction of mycophenolate mofetil, azathioprine may be relegated to a second-line antimetabolite for the prevention of graft rejection.

Cyclophosphamide

Immuno-suppression

Antimicrobials

Cardiovascular Agents

Antiosteo-porosis Agents

Antiplatelets

Diabetes Agents

Ulcer Treatment

Diuretics

Other Agents

Brand Name	**Cytoxan®**
Company	Bristol-Myers Oncology Division
Class	• Immunosuppressive antimetabolite
Mechanism of Action	• Interferes with the growth of susceptible, rapidly proliferating cells, possibly through cross-linking of cellular DNA
Application in Transplantation	• Adjunct to other immunosuppressive therapy to delay onset of first rejection episode
Contraindications	• Severe myelosuppression • Hypersensitivity to cyclophosphamide
Warnings	• Risk of myeloproliferative and lymphoproliferative malignancies • Fetal damage • Sterility in both sexes • Amenorrhea • Oligospermia or azoospermia • Hemorrhagic cystitis • Urinary bladder fibrosis • Cardiotoxicity at high doses (120 mg/kg to 270 mg/kg) • Risk of infection

Special Precautions	• Treatment should be discontinued if any of the following are present:
	  • leukopenia
	  • thrombocytopenia
	  • previous exposure to other cytotoxic agents
	  • impaired renal function
	  • impaired hepatic function
	• Hematologic profile and urine should be monitored periodically
	• Should be used with caution in adrenalectomized patients
	• Pregnancy Category D

Adverse Reactions	• Sterility in both sexes • Oligospermia or azoospermia • Nausea and vomiting • Leukopenia • Alopecia • Hemorrhagic cystitis • Urinary bladder fibrosis • Renal tubular necrosis • Opportunistic infection
Drug Interactions	• Phenobarbitol • Succinylcholine • Anesthesiologist should be informed if patient is to receive general anesthesia within 10 days of treatment
Formulations	*Cytoxan for Injection* • 100 mg vial • 200 mg vial • 500 mg vial • 1.0 g vial • 2.0 g vial *Lyophilized Cytoxan for Injection* • 100 mg vial • 200 mg vial • 500 mg vial • 1.0 g vial • 2.0 g vial
Dosage	• 2 mg/kg/d to 3 mg/kg/d is initial recommended dose, but is rapidly reduced due to toxicity

Editors' Notes:
Cyclophosphamide is given in patients with contraindications to azathioprine. The dose is usually half of the azathioprine dose. Prolonged use of cyclophosphamide is not recommended due to the possibility of future malignancy.

Cyclosporine-A (non-modified)

Brand Name	Sandimmune®	Cyclosporine
Company	Novartis[1]	Torpharm[2]
Class	• Immunosuppressant produced as a metabolite by the fungus species *Beauvaria nivea* Gams[1]; *Tolyprocladium inflatum* Gams[2]	
Mechanism of Action	• Preferential inhibition of T lymphocytes • Suppresses activation of T lymphocytes by inhibiting production and release of lymphokines, specifically interleukin-2	
Indications	• Prophylaxis of graft rejection in kidney, liver, and heart allogeneic transplantation • Treatment of chronic rejection previously treated with other immuno-suppressants	
Contraindication	• Hypersensitivity to cyclosporine or polyoxyethylated castor oil	
Warnings	• Nephrotoxicity • Hepatotoxicity • Increased susceptibility to infection and lymphoma • Erratic absorption of soft gelatin capsules and oral solution necessitates repeated monitoring of cyclosporine blood levels • Anaphylactic reactions with IV formulation	
Special Precautions	• Hypertension may occur and require therapy with antihypertensives (potassium-sparing diuretics should not be used) • Repeated laboratory monitoring is required • Pregnancy Category C	
Adverse Reactions	• Renal dysfunction (25% - renal, 38% - cardiac, 37% - hepatic) • Tremor (21% - renal, 31% - cardiac, 55% - hepatic) • Hirsutism (21% - renal, 28% - cardiac, 45% - hepatic) • Hypertension (13% - renal, 53% - cardiac, 27% - hepatic) • Gum hyperplasia (9% - renal, 5% - cardiac, 16% - hepatic)	
Drug Interactions	*Drugs with Synergistic Nephrotoxicity* • Gentamicin • Amphotericin B • Tobramycin • Ketoconazole • Vancomycin • Melphalan • Cimetidine • TMP-SMX • Ranitidine • Azapropazon • Diclofenac • Naproxen • Sulindac *Drugs that Increase Cyclosporine Levels* • Diltiazem • Ketoconazole • Nicardipine • Erythromycin • Verapamil • Itraconazole • Danazol • Bromocriptine • Fluconazole • Methylprednisolone • Metoclopramide • Drugs which inhibit cytochrome P450 3A4 *Drugs that Decrease Cyclosporine Levels* • Rifampin • Phenobarbitol • Carbamazepine • Phenytoin • Drugs which induce cytochrome P450 3A4 *Reduced Clearance with Cyclosporine* • Prednisone • Lovastatin • Digoxin • Grapefruit juice can increase cyclosporine concentrations *Other* • Vaccinations—live vaccinations should be avoided	

Continued ...

Cyclosporine-A (non-modified)

Formulations	• 25 mg and 100 mg soft gelatin capsules • 50 mL bottle containing 100 mg/mL for oral solution • 5 mL vial containing 50 mg/mL
Dosage	*Gelatin Capsules and Oral Solution* • 15 mg/kg single dose given 4 to 12 hours prior to transplantation • Single daily dose is continued postoperatively for 1 to 2 weeks and then tapered by 5% each week until maintenance dose of 5 mg/kg/d to 10 mg/kg/d is reached *IV Infusion* • 5 mg/kg to 6 mg/kg single dose given 4 to 12 hours prior to transplantation • Single daily IV dose continued postoperatively until patient can tolerate oral formulations

Editors' Notes:

Sandimmune® is the original formulation of cyclosporine used in clinical transplantation. Erratic absorption of the formulation has limited its use in transplant recipients. Many long-term transplant recipients continue on Sandimmune®. The formulation has been supplanted by the micro-emulsion formula of cyclosporine.

Antimicrobials

Cardiovascular Agents

Antiosteo-porosis Agents

Antiplatelets

Diabetes Agents

Ulcer Treatment

Diuretics

Other Agents

Cyclosporine Capsules and Oral Solution (modified)

Antimicrobials

Cardiovascular Agents

Antiosteo-porosis Agents

Antiplatelets

Diabetes Agents

Ulcer Treatment

Diuretics

Other Agents

Brand Name	Neoral®	Gengraf		Cyclosporine	Cyclosporine
Company	Novartis[1]	Abbot Laboratories/SangStat[2]		Eon[3]	Pliva[4]
Class	• Immunosuppressant produced as a metabolite by the fungus species *Beauvaria nivea* Gams[1]; *Aphanocladium album*[2]; *Cordyceps militaris*[3,4]				
Mechanism of Action	• Preferential inhibition of T lymphocytes • Suppresses activation of T lymphocytes by inhibiting production and release of lymphokines, specifically interleukin-2				
Indications	• Prophylaxis of graft rejection in kidney, liver, and heart allogeneic transplantation • Rheumatoid arthritis • Psoriasis				
Contraindication	• Hypersensitivity to cyclosporine or any of the ingredients of the formulation				
Warnings	• Nephrotoxicity • Hepatotoxicity • Increased susceptibility to infection and lymphoma • Reports of convulsions in pediatric and adult patients, especially when used in conjunction with methylprednisolone • Modified cyclosporine is not bioequivalent to Sandimmune				
Special Precautions	• Any change in cyclosporine formulation should be made cautiously and under the advisement of a physician—Patients should be given detailed dosage instructions • Hypertension may occur and require therapy with antihypertensives (potassium-sparing diuretics should not be used) • Repeated laboratory monitoring is required • Pregnancy Category C				
Adverse Reactions	• Renal dysfunction • Hyperkalemia • Tremor • Hyperurecemia • Hirsutism • Encephalopathy • Hypertension • Gum hyperplasia				
Drug Interactions	*Drugs with Synergistic Nephrotoxicity* • Gentamicin • Amphotericin B • Naproxen • Tobramycin • Ketoconazole • Sulindac • Vancomycin • Melphalan • Colchicine • TMP-SMX • Tacrolimus • Non-steroidal • Cimetidine • Diclofenac anti-inflammatory • Ranitidine • Azapropazon agents *Drugs that Increase Cyclosporine Levels* • Diltiazem • Clarithromycin • Nicardipine • Allopurinol • Verapamil • Danazol • Ketoconazole • Bromocriptine • Fluconazole • Methylprednisolone • Itraconazole • Metoclopramide • Erythromycin • Drugs which inhibit cytochrome P450 3A4 • Quinupristin/Dalfopristin • Colchicine • Amiodarone • HIV-protease inhibitors				

Continued ..

Cyclosporine Capsules and Oral Solution (modified)

Drug Interactions (continued)	*Drug/Dietary Supplements that Decrease Cyclosporine Levels* • Rifampin • Carbamazepine • Nafcillin • Octreotide • Phenytoin • Ticlopidine • Phenobarbitol • Drugs which induce cytochrome P450 3A4 • St. John's Wort • Orlistat *Reduced Clearance with Cyclosporine* • Prednisone • Lovastatin • Digoxin • Grapefruit and grapefruit juice can increase cyclosporine concentrations *Other* • Vaccinations—live vaccinations should be avoided
Formulations	• Neoral 25 mg and 100 mg soft gelatin capsules in 30-count blister packages • Oral solution 100 mg/mL in bottle containing 50 mL • Gengraf 25 mg and 100 mg capsules in 30-count unit dose packages • Eon cyclosporine 25 mg and 100 mg capsules in 30-count unit dose blisters • Pliva cyclosporine 25 mg and 100 mg capsules in 30-count unit dose blisters
Dosage	**NOTE: Neoral and Sandimmune are not bioequivalent and should not be used interchangeably. Neoral has increased bioavailability and this should be taken into consideration when making dosing decisions.** • Daily dose should be given as two divided doses on a consistent schedule *Newly Transplanted Patients* • Initial dose of Neoral should be the same as a Sandimmune dose. Suggested initial doses include: – 9 ± 3 mg/kg/day for kidney transplant patients – 8 ± 4 mg/kg/day for liver transplant patients – 7 ± 3 mg/kg/day for heart transplant patients • The dose is then subsequently adjusted to achieve a predefined cyclosporine blood concentration *Conversion from Sandimmune to Neoral* • Neoral should be started with the same daily dose as was previously used with Sandimmune (1:1 dose conversion) • The Neoral dose should then be adjusted to achieve preconversion cyclosporine blood trough concentrations • Until cyclosporine blood trough concentrations reach preconversion levels, monitoring should be undertaken every 4 to 7 days

Antimicrobials

Cardiovascular Agents

Antiosteo-porosis Agents

Antiplatelets

Diabetes Agents

Ulcer Treatment

Diuretics

Other Agents

Editors' Notes:

Cyclosporine has been used in clinical transplantation for nearly 20 years. However, it is not clear that trough monitoring of CSA is the best way to monitor for adequate immunosuppression. Recently, C2 monitoring is being advocated as better than trough monitoring for better efficacy and preventing toxicity. The CSA blood level at 2 hours (C2) is a more consistent and reliable measure of the absorption of cyclosporine. (Cyclosporine microemulsion (Neoral) absorption profiling and sparse-sample predictors during the first 3 months after renal transplantation. ©2002 Am J Transplant.)

The use of cyclosporine maintenance therapy has declined. In 2004, 19% of transplant patients were receiving cyclosporine maintenance. (Meier-Kriesche et al. AJT 2006; 6(Part 2):1111-1131.)

Methylprednisolone

Brand Name	**Solu-medrol®**
Company	Pfizer
Class	• Anti-inflammatory steroid
Mechanism of Action	• Causes emigration of circulating T cells from intravascular tissue compartment to lymphoid tissue • Inhibits production of T-cell lymphokines that are needed to amplify macrophage and lymphocyte response
Applications in Transplantation	• Immunosuppressive adjunct for the prevention and treatment of solid organ rejection • Attenuation of cytokine release syndrome in patients treated with muromonab-CD3
Contraindications	• Hypersensitivity to methylprednisolone and/or its components • Systemic fungal infections • Prematurity in infants
Warnings	• May produce posterior subcapsular cataracts, glaucoma, and may enhance the establishment of secondary ocular infections due to fungi or viruses • Increased calcium excretion • Vaccination should not be undertaken during therapy • Cardiac arrhythmias, circulatory collapse, and/or cardiac arrest have occurred after rapid administration • Adequate human reproduction studies have not been undertaken
Special Precautions	• Use with caution in patients with hypothyroidism, cirrhosis, ocular herpes simplex, hypertension, congestive heart failure, and ulcerative colitis • Psychologic derangements may occur while on therapy
Adverse Reactions	• Sodium and fluid retention, congestive heart failure in susceptible patients, potassium loss, hypokalemic alkalosis, hypertension • Muscle weakness, steroid myopathy, loss of muscle mass, severe arthralgia, vertebral compression fractures, aseptic necrosis of humeral and femoral heads, pathologic fracture of long bones, osteoporosis • Peptic ulcer with possible perforation and hemorrhage, pancreatitis, abdominal distension, ulcerative esophagitis • Impaired wound healing, thin fragile skin, petechiae and ecchymoses, facial erythema, increased sweating, suppressed reaction to skin tests • Pseudotumor cerebri, convulsions, vertigo, headache • Cushingoid state, suppression of growth in children, secondary adrenocortical and pituitary unresponsiveness, menstrual irregularities, decreased carbohydrate tolerance, manifestations of latent diabetes mellitus, increased requirements for insulin or oral hypoglycemic agent • Posterior subcapsular cataracts, increased intraocular pressure, glaucoma, exophthalmos
Drug Interactions	• Barbiturates • Phenytoin • Rifampin • Salicylates • Vaccines • Toxoids
Formulations	• 40 mg single dose vial • 125 mg single dose vial • 500 mg vial • 500 mg vial with diluent • 1 g vial • 1 g vial with diluent

Continued …

Methylprednisolone

Dosage	Induction • 250 mg to 1000 mg at time of transplantation and for next 2 to 3 doses *Taper* • Start at 2 mg/kg/d, taper to a range of 0.15 mg/kg/d to 0.2 mg/kg/d after one year *Attenuation of Cytokine Release Syndrome* • 8 mg/kg given 1 to 4 hours prior to first injection of muromonab-CD3

Editors' Notes:

Multiple different regimens of methylprednisolone have been utilized to treat rejection. The initial higher doses are the most important in controlling rejection. It is unclear whether prolonged tapering of steroids following rejection is of value.

Antimicrobials

Cardiovascular Agents

Antiosteo-porosis Agents

Antiplatelets

Diabetes Agents

Ulcer Treatment

Diuretics

Other Agents

Mycophenolate Mofetil

Brand Name	CellCept®
Company	Roche Laboratories
Class	• Immunosuppressive antimetabolite
Mechanism of Action	• Selectively inhibits inosine monophosphate dehydrogenase in the de novo pathway of purine synthesis, producing potent cytostatic effects on T and B lymphocytes
Indication	• Prophylaxis of graft rejection in patients receiving renal and cardiac allogeneic transplantation
Contraindication	• Hypersensitivity to mycophenolate mofetil, mycophenolic acid or any components of the drug
Warnings	• Increased susceptibility to infection and lymphoma • Adverse effects on fetal development have been observed in pregnant rats and rabbits—mycophenolate mofetil should not be used in pregnant women and contraception should be used during therapy • Neutropenia has been observed
Special Precaution	• Gastrointestinal hemorrhage may occur • Patients with renal impairment have shown higher MPA and MPAG AUCs than normal volunteers • Should not be used in conjunction with azathioprine • Repeated laboratory monitoring is required • Pregnancy Category C
Adverse Reactions	• Diarrhea • Leukopenia • Sepsis • Vomiting
Drug Interactions	• Acyclovir • Antacids with magnesium and aluminum hydroxides • Cholestyramine • Drugs that alter gastrointestinal flora may interact with mycophenolate mofetil by disrupting enterohepatic recirculation • Probenecid • Oral contraceptives
Formulation	• 250 mg capsules supplied in bottles of 100 and 500 • 500 mg tablets in bottles of 100 and 500 • CellCept intravenous: 20 mL, sterile vial containing 500 mg mycophenolate mofetil
Dosage	• 1 gram twice a day used in combination with corticosteroids and cyclosporine • Initial dose should be given within 72 hours following transplantation

Antimicrobials

Cardiovascular Agents

Antiosteo-porosis Agents

Antiplatelets

Diabetes Agents

Ulcer Treatment

Diuretics

Other Agents

Editors' Notes:

Mycophenolate mofetil reduces the risk of first acute rejection by 50%. Toxicity is minor, but includes bone marrow suppression and gastrointestinal complaints. A higher incidence of CMV disease compared to azathioprine control was observed in the clinical trials. Recent registry studies appear to indicate that renal transplant recipients receiving mycophenolate have improved long-term outcomes.

The metabolism of mycophenolate is altered by coadministration with cyclosporine. Mycophenolic acid (MPA) levels are lower when mycophenolate mofetil is given with cyclosporine. Lower doses should be considered in recipients receiving tacrolimus or steroids alone without a calcineurin inhibitor.

Mycophenolate Sodium Delayed-Release

Brand Name	Myfortic®
Company	Novartis
Class	• Immunosuppressive antimetabolite
Mechanism of Action	• Selectively inhibits inosine monophosphate dehydrogenase in the de novo pathway of purine (guanosine) synthesis, producing potent cytostatic effects on T and B lymphocytes
Indication	• Prophylaxis of organ rejection in patients receiving allogeneic renal transplants, administered in combination with cyclosporine and corticosteroids
Contraindications	• Hypersensitivity to mycophenolate sodium, mycophenolic acid, mycophenolate mofetil, or to any components of the drug
Warnings	• Increased susceptibility to infection and lymphoma and other neoplasms • Adverse effects on fetal development have been observed in pregnant rats and rabbits —mycophenolate sodium should not be used in pregnant women and contraception should be used during and for 6 weeks after therapy • Neutropenia has been observed
Special Precautions	• Gastrointestinal hemorrhage may occur • Patients with renal impairment have shown to have higher MPA and MPAG AUCs than normal volunteers • Live attenuate vaccines should be avoided • Should not be used in conjunction with azathioprine • Regular laboratory monitoring is required • Pregnancy Category C
Adverse Reactions	• Nausea/vomiting/diarrhea • Leukopenia • Sepsis
Drug Interactions	• Acyclovir/ganciclovir • Magnesium-aluminum containing antacids • Bile acid sequestrates: cholestyramine/ colesevelam • Agents that may interfere with enterohepatic recirculation • Oral contraceptives • Probenecid
Formulation	• 360 mg tablets in bottles of 120 • 180 mg tablets in bottles of 120
Dosage	• 720 mg orally twice daily used in combination with other immuno-suppressants

Antimicrobials

Cardiovascular Agents

Antiosteo-porosis Agents

Antiplatelets

Diabetes Agents

Ulcer Treatment

Diuretics

Other Agents

Editors' Notes:

There are no significant clinical differences comparing Myfortic® to CellCept® in clinical trials (Salvadori M et al. Am J Transplant 2004 Feb; 4(2):231-236.)

One potential advantage of the Myfortic® formulation is less gastrointestinal intolerance and dose adjustments. This is the subject of a randomized trial which is underway.

Prednisone

Brand Name	Deltasone®
Company	Pharmacia and Upjohn, Inc.
Class	• Anti-inflammatory steroid
Mechanism of Action	• Causes emigration of circulating T cells from intravascular tissue compartment to lymphoid tissue • Inhibits production of T cell lymphokines that are needed to amplify macrophage and lymphocyte response
Application in Transplantation	• Immunosuppressive adjunct for the prevention and treatment of solid organ rejection
Contraindications	• Hypersensitivity to prednisone and/or its components • Systemic fungal infection
Warnings	• May produce posterior subcapsular cataracts, glaucoma, and may enhance the establishment of secondary ocular infections due to fungi or viruses • Increased calcium excretion • Vaccination should not be undertaken during therapy • Adequate human reproduction studies have not been undertaken
Special Precautions	• Use with caution in patients with hypothyroidism, cirrhosis, ocular herpes simplex, hypertension, congestive heart failure, and ulcerative colitis • Psychologic derangements may occur while on therapy
Adverse Reactions	• Sodium and fluid retention, congestive heart failure in susceptible patients, potassium loss, hypokalemic alkalosis, hypertension • Muscle weakness, steroid myopathy, loss of muscle mass, vertebral compression fractures, aseptic necrosis of humeral and femoral heads, pathologic fracture of long bones, osteoporosis • Peptic ulcer with possible perforation and hemorrhage, pancreatitis, abdominal distension, ulcerative esophagitis • Impaired wound healing, thin fragile skin, petechiae and ecchymoses, facial erythema, increased sweating, suppressed reaction to skin tests • Pseudotumor cerebri, convulsions, vertigo, headache • Cushingoid state, suppression of growth in children, secondary adrenocortical and pituitary unresponsiveness, menstrual irregularities, decreased carbohydrate tolerance, manifestations of latent diabetes mellitus, increased requirements for insulin or oral hypoglycemic agent • Posterior subcapsular cataracts, increased intraocular pressure, glaucoma, exophthalmos
Drug Interactions	• Barbiturates • Phenytoin • Rifampin • Salicylates • Vaccines • Toxoids
Dosage Strengths	• 2.5 mg tablets • 5 mg tablets • 10 mg tablets • 20 mg tablets • 50 mg tablets
Dosage	*Maintenance—Adults* • 0.1 mg/kg/d to 2 mg/kg/d usually given once daily *Maintenance—Pediatric* • 0.25 mg/kg/d to 2 mg/kg/d or 25 mg/m² to 60 mg/m² usually given daily or on alternate days

Editors' Notes:

Steroid avoidance or early steroid withdrawal is under active investigation in many transplant centers. The short-term results are acceptable; long-term efficacy is yet to be established.

In 2004, 23% of first transplants were discharged without steroids. (Meier-Kriesche et al. AJT 2006; 6(Part 2):1111-1131.)

Tacrolimus

Brand Name	Prograf®
Company	Astellas
Class	• Macrolide antibiotic with immunosuppressant properties
Mechanism of Action	• Binds to a T-cell binding protein and prevents synthesis of interleukin-2 and other lymphokines essential to T-lymphocyte function
Indications	• Prophylaxis of graft rejection in liver and kidney allogeneic transplantation • It is recommended that tacrolimus be used concomitantly with adrenal corticosteroids
Contraindications	• Hypersensitivity to tacrolimus • Hypersensitivity to HCO-60 (polyoxyl 60 hydrogenated castor oil) with IV formulation
Warnings	• Increased incidence of posttransplant diabetes mellitus and insulin use at 24 months in kidney transplant recipients • Neurotoxicity • Nephrotoxicity • Hyperkalemia • Increased risk of infection and lymphomas • Patients should be monitored closely for at least the first 30 minutes of therapy for signs of anaphylactic reactions
Special Precautions	• Hypertension is a common occurrence with tacrolimus and may require treatment with antihypertensive agents. Since tacrolimus may cause hyperkalemia, potassium-sparing diuretics should be avoided • Hyperglycemia may occur and require treatment • Lower doses should be used in patients with renal insufficiencies • Patients with hepatic impairment may have a higher risk of developing renal insufficiency • Patients should be informed of the need for regular laboratory monitoring • Myocardial hypertrophy • Pregnancy Category C
Adverse Reactions	• Tremor • Nausea • Headache • Renal dysfunction • Diarrhea • Paresthesia • Hypertension • Hypomagnesemia
Drug Interactions	*Drugs with Synergistic Nephrotoxicity* • Gentamicin • Amphotericin B • Tobramycin • Ketoconazole • Vancomycin • Melphalan • TMP-SMX • Diclofenac • Cimetidine • Azapropazon • Ranitidine *Drugs that Increase Tacrolimus Levels* • Diltiazem • Clarithromycin • Nicardipine • Cimetidine • Verapamil • Danazol • Ketoconazole • Bromocriptine • Fluconazole • Methylprednisolone • Itraconazole • Metoclopramide • Erythromycin • Cyclosporine

Antimicrobials

Cardiovascular Agents

Antiosteo-porosis Agents

Antiplatelets

Diabetes Agents

Ulcer Treatment

Diuretics

Other Agents

Continued …

Tacrolimus

Drug Interactions (continued)	*Drugs that Increase Tacrolimus Levels* • Nifedipine • Ketoconazole • Ethinyl estradiol • Nefazodone • Grapefruit and grapefruit juice can increase tacrolimus levels • Clotrimazole • Troleandomycin • Omeprazole • Protease inhibitors • Drugs that inhibit cytochrome P450 3A4 *Drugs/Dietary Supplements that Decrease Tacrolimus Levels* • Rifampin • Rifabutin • Phenytoin • Drugs that induce cytochrome P450 3A4 • Carbamazepine • Phenobarbitol • St. John's Wort *Other* • Vaccinations—live vaccinations should be avoided
Formulations	• 0.5 mg capsules • 1 mg capsules • 5 mg capsules • 1 mL ampules containing the equivalent of 5 mg of anhydrous tacrolimus per mL—supplied in boxes of 10 ampules
Dosage and Administration	*IV Infusion* • 0.03 to 0.05 mg/kg/d as a continuous infusion • Patients should be converted to oral therapy as soon as can be tolerated *Capsules* • Liver: 0.10 to 0.15 mg/kg/d • Kidney: 0.2 mg/kg/d • First dose should be given 8 to 12 hours after discontinuing IV infusion

Editors' Notes:

Tacrolimus appears to be a more effective drug than cyclosporine for liver transplantation. Rescue for refractory liver and kidney rejection has also been shown to be an important use for the drug. A double-blind study comparing Sandimmune and FK506 in cadaveric renal transplantation demonstrated superior efficacy of FK506 for rejection prophylaxis. A higher incidence of PTDM was noted, but was reversible in 50% of patients at 2 years. The 5-year data from this study was recently published. Treatment failure was significantly less frequent in tacrolimus-treated recipients (43.8% vs. 56.8%; p=0.008). With cross-over due to rejection counted as graft failure, graft survival was significantly better with tacrolimus (63.8% vs. 53.8%; p=0.014). Hypertension and hyperlipidemia were less with tacrolimus. Nearly 50% of patients who required insulin with tacrolimus were able to discontinue insulin. (Vincenti F, Jensik SC, Filo RS et al. A long-term comparison of tacrolimus (FK-506) and cyclosporine in kidney transplantation: evidence for improved allograft survival at five years. ©2002 Transplantation. Lippincott, Williams and Wilkins.)

Tacrolimus is the most commonly used calcineurin inhibitor in 60% of patients discharged in 2004. (Meier-Kriesche et al. AJT 2006; 6(Part 2):1111-1131.)

Avoid taking with meals, antacids and divalent cation supplements.

Sirolimus

Brand Name	**Rapamune®**
Company	Wyeth-Ayerst
Class	• Macrocyclic lactone antibiotic with immunosuppressant properties
Mechanism of Action	• Binds to an immunophilin protein to form a complex which inhibits the activation of the mammalian Target of Rapamycin (mTOR), a regulatory kinase. This inhibits T-lymphocyte activation and proliferation by IL-2, IL-4, and IL-5
Indications	• The prophylaxis of organ rejection in patients receiving renal transplants
Contraindications	• Hypersensitivity to sirolimus or its derivatives or any component of the drug product
Warnings	• Increased risk of infection and lymphomas
Special Precautions	• Increased serum cholesterol and triglycerides • Lymphocele • Impaired renal function in combination with cyclosporine • Pregnancy Category C
Adverse Reactions	• Hypertension • Rash • Acne • Anemia • Arthralgia • Diarrhea • Hypokalemia • Thrombocytopenia • Leukopenia • Fever
Drug Interactions	• Sirolimus is known to be a substrate for both cytochrome CYP3A4 and P-glycoprotein. *Drugs that Increase Sirolimus Levels* • Cyclosporine (amount affected by coadministration schedule and formulation) • Diltiazem • Ketoconazole • Rifampin *Drugs that May Increase Sirolimus Levels Include:* • Calcium channel blockers: nicardipine, verapamil • Antifungal agents: clotrimazole, fluconazole, itraconazole • Macrolide antibiotics: clarithromycin, erythromycin, troleandomycin • Gastrointestinal prokinetic agents: cisapride, metoclopramide • Other drugs: bromocriptine, cimetidine, danazol, HIV-protease inhibitors (e.g., ritonavir, indinavir) • Grapefruit and grapefruit juice may increase sirolimus concentrations. *Drugs that May Decrease Sirolimus Levels Include:* • Anticonvulsants: carbamazepine, phenobarbital, phenytoin • Antibiotics: rifabutin, rifapentine • Herbal preparations: St. John's Wort (hypericum perforatum) could result in reduced sirolimus levels.

Continued ..

Immuno-suppression

Antimicrobials

Cardiovascular Agents

Antiosteo-porosis Agents

Antiplatelets

Diabetes Agents

Ulcer Treatment

Diuretics

Other Agents

Sirolimus

Formulations	• Oral solution in a concentration of 1 mg/mL in: 2 oz (60 mL fill) amber glass bottles • 1 mg tablets supplied in a bottle of 100 tablets • 1 mg tablets supplied in a carton of 100 tablets (10 blister cards of 10 tablets each) • 2 mg tablets supplied in a bottle of 100 tablets • 2 mg tablets supplied in a carton of 100 tablets (10 blister cards of 10 tablets each)
Dosage	• De novo transplant recipients, a loading dose of sirolimus of 3 times the maintenance dose should be given. A daily maintenance dose of 2 mg is recommended for use in renal transplant patients, with a loading dose of 6mg. • The initial dose in patients >12 years old, who weigh less than 40 kg the loading dose should be 3 mg/m² followed by 1 mg/m²/day.

Editors' Notes:

To minimize the variability of blood concentrations, sirolimus should be taken consistently in relation to time of administration with or without cyclosporine and/or food.

Sirolimus solution—For simultaneous administration, the mean C_{max} and AUC of sirolimus were increased by 116% and 230%, respectively, relative to administration of sirolimus alone. However, when given 4 hours after Neoral® Soft Gelatin Capsules (cyclosporine capsules [MODIFIED]) administration, sirolimus C_{max} and AUC were increased by 37% and 80%, respectively, compared to sirolimus alone. Cyclosporine clearance was reduced only after multiple-dose administration over 6 months.

Sirolimus tablets—For simultaneous administration, mean C_{max} and AUC were increased by 512% and 148%, respectively, relative to administration of sirolimus alone. However, when given 4 hours after cyclosporine administration, sirolimus C_{max} and AUC were both increased by only 33% compared with administration of sirolimus alone.

Sirolimus was used in only 9% of patients discharged after kidney transplantation in 2004. (Meier-Kriesche et al. AJT 2006; 6(Part 2):1111-1131.)

Sirolimus has also been associated with worsening proteinuria following conversion from a calcineurin inhibitor. (van den Akker et al. Kidney Int 2006; 70(7):1355-1357.)

Antimicrobials

Cardiovascular Agents

Antiosteo-porosis Agents

Antiplatelets

Diabetes Agents

Ulcer Treatment

Diuretics

Other Agents

Chapter 2
Antimicrobials

Antivirals Overview
- Acyclovir
- Valacyclovir
- Ganciclovir
- Valganciclovir
- Cytomegalovirus (CMV) Immune Globulin
- Foscarnet Sodium
- Cidofovir
- Lamivudine
- Adefovir Dipivoxil
- Oseltamivir Phosphate

Antibiotics Overview
- Trimethoprim and Sulfamethoxazole (TMP-SMX)
- Penicillins with Beta-Lactamase Inhibitor
- Quinolones
- Cephalosporins
- Aminoglycosides
- Linezolid
- Tigecycline
- Daptomycin

Antifungals Overview
- Nystatin Suspension
- Clotrimazole Troche
- Fluconazole
- Posaconazole
- Voriconazole
- Amphotericin B
- Amphotericin B Lipid Complex Injection
- Miconazole Vaginal Cream 2%
- Ketoconazole
- Caspofungin
- Micafungin
- Anidulafungin

Antivirals Overview

Agent	Dosage
Acyclovir	*Mucosal and Cutaneous HSV Infection in Immunocompromised Patients* • In adults, 5 mg/kg IV infused at a constant rate over 1 hour, every 8 hours (15 mg/kg/d IV) for 7 days • In children under 12 years of age, 10 mg/kg IV at a constant rate over 1 hour, every 8 hours for 7 days *HSV Encephalitis* • In adults, 10 mg/kg IV infused at a constant rate over 1 hour, every 8 hours for 10 days • In children between 3 months and 12 years of age, 20 mg/kg IV at a constant rate over 1 hour, every 8 hours for 10 days *Varicella Zoster Infection in Immunocompromised Patients* • In adults, 10 mg/kg IV infused at a constant rate over 1 hour, every 8 hours for 7 days • In children under 12 years of age, 20 mg/kg IV at a constant rate over 1 hour, every 8 hours for 7 days
Valacyclovir	*Herpes Zoster* • 1 gram orally 3 times daily for 7 days *Genital Herpes* • Initial episodes: 1 gram twice daily for 3 days • Recurrent episodes: 500 mg twice daily for 3 days • Suppressive therapy: 0.5 - 1 gram once daily
Ganciclovir	*Prevention and Treatment of CMV in Transplant* • 5 mg/kg IV given at a constant rate over 1 hour every 12 hours for 14 to 21 days • Followed by 5 mg/kg/d IV daily, or 6 mg/kg/d IV 5 days per week, or 1000 mg orally three times daily • Duration of therapy depends on degree of immunosuppression; in clinical trials, treatment was continued until postop day 100 to 120
Valganciclovir	• Induction phase: 900 mg orally twice daily • Maintenance phase: 900 mg once daily
Cytomegalovirus (CMV) Immune Globulin	• 150 mg/kg within 72 hours of transplantation • 100 mg/kg at 2, 4, 6, and 8 weeks posttransplant • 50 mg/kg at 12 and 16 weeks posttransplant
Foscarnet Sodium	*Induction—Normal Renal Function* • 60 mg/kg IV at a constant rate over a minimum of 1 hour every 8 hours for 2 to 3 weeks depending on clinical response *Maintenance—Normal Renal Function* • 90 mg/kg/d to 120 mg/kg/d IV infused over 2 hours

Continued ...

Immuno-suppression · Antimicrobials · Cardiovascular Agents · Antiosteoporosis Agents · Antiplatelets · Diabetes Agents · Ulcer Treatment · Diuretics · Other Agents

Antivirals Overview

Agent	Dosage
Cidofovir	*CMV Retinitis in AIDS patients* • Induction phase—5 mg/kg IV once weekly for two weeks • Maintenance phase—5 mg/kg IV once every other week until retinitis progression or therapy-limiting toxicity *BK Nephropathy* • 0.25-1 mg/kg IV every other week (probenecid should not be given)
Lamivudine	*Chronic Hepatitis B Infection* • 100 mg orally once daily
Adefovir Dipivoxil	*Chronic Hepatitis B Infection* • 10 mg orally once daily (optimal duration of therapy unknown) without regard to food
Oseltamivir Phosphate	*Chronic Hepatitis B Infection* • 75 mg orally twice daily for 5 days

Antimicrobials

Cardiovascular
Agents

Antiosteo-
porosis Agents

Antiplatelets

Diabetes
Agents

Ulcer
Treatment

Diuretics

Other
Agents

Acyclovir

Immuno-suppression

Antimicrobials

Cardiovascular Agents

Antiosteo-porosis Agents

Antiplatelets

Diabetes Agents

Ulcer Treatment

Diuretics

Other Agents

Brand Name	Zovirax®
Company	Glaxo Wellcome Inc.
Class	Synthetic nucleoside analogue with antiviral properties against herpesviruses
Mechanism of Action	• Interferes with DNA polymerase of herpesvirus cells, inhibiting their replication
Indications	• PO - acute treatment of herpes zoster (shingles) • Initial episodes and the management of genital herpes • Treatment of chickenpox (varicella) • IV - treatment of initial and recurrent mucosal and cutaneous herpes simplex (HSV-1 and HSV-2) in immunocompromised patients • Treatment of herpes simplex encephalitis • Treatment of neonatal herpes infections • Treatment of varicella-zoster (shingles) infections in immunocompromised patients
Contraindication	• Hypersensitivity to acyclovir or valacyclovir
Warning	• Acyclovir sodium is intended only for intravenous infusion, which should be given over a period of 1 hour to reduce the risk of renal tubular damage • Thrombotic thrombocytopenic purpura/hemolytic uremic syndrome (TTP/HUS), which has resulted in death, has occurred in immunocompromised patients receiving acyclovir therapy.
Special Precautions	• Recommended dosage, frequency, and length of treatment should not be exceeded • Precipitation of acyclovir crystals in renal tubules can occur, potentially leading to renal tubular damage • Dosage adjustments should be based on estimated creatinine clearance • 1% of patients have experienced encephalopathic changes • Resistant strains have developed after repeated exposure to acyclovir • Pregnancy Category B
Adverse Reactions	• Inflammation or phlebitis at injection site • Transient elevations in serum creatinine and BUN • Nausea and/or vomiting • Itching, rash, or hives • Elevation of transaminases • Encephalopathic changes, including lethargy, obtundation, tremors, confusion, hallucinations, agitation, seizures, and coma
Drug Interaction	• Probenecid
Formulations	• 10 mL vial containing 500 mg • 20 mL vial containing 1000 mg • Also available in ointment, capsules, tablets, and suspension formulations
Dosage	Caution: Rapid or bolus intravenous and intramuscular or subcutaneous injection must be avoided

Continued ...

Immuno-suppression

Antimicrobials

Cardiovascular Agents

Antiosteo-porosis Agents

Antiplatelets

Diabetes Agents

Ulcer Treatment

Diuretics

Other Agents

Acyclovir

| **Dosage** | **Intravenous:** |

Mucosal and Cutaneous Herpes Simplex (HSV-1 and HSV-2) Infections in Immunocompromised Patients
- Adults and adolescents (12 years of age and older): 5 mg/kg infused at a constant rate over 1 hour, every 8 hours for 7 days
- Pediatrics (under 12 years of age): 10 mg/kg infused at a constant rate over 1 hour, every 8 hours for 7 days

Severe Initial Clinical Episodes of Herpes Genitalis
- Adults and adolescents (12 years of age and older): 5 mg/kg infused at a constant rate over 1 hour, every 8 hours for 5 days

Herpes Simplex Encephalitis
- Adults and adolescents (12 years of age and older): 10 mg/kg infused at a constant rate over 1 hour, every 8 hours for 10 days
- Pediatrics (3 months to 12 years of age): 20 mg/kg infused at a constant rate over 1 hour, every 8 hours for 10 days

Neonatal Herpes Simplex Virus Infections (Birth to 3 months)
- 10 mg/kg infused at a constant rate over 1 hour, every 8 hours for 10 days. In neonatal herpes simplex infections, doses of 15 mg/kg or 20 mg/kg (infused at a constant rate over 1 hour every 8 hours) have been used; the safety and efficacy of these doses is not known.

Herpes Zoster in Immunocompromised Patients
- Adults and adolescents (12 years of age and older): 10 mg/kg infused at a constant rate over 1 hour, every 8 hours for 7 days
- Pediatrics (under 12 years of age): 20 mg/kg infused at a constant rate over 1 hour, every 8 hours for 7 days

Obese patients should be dosed at the recommended adult dose using the Ideal Body Weight

Dosing Adjustments for Patients with Renal Impairment

Creatinine clearance (mL/min)	% of recommended dose	Dosing interval (h)
>50	100%	8
25 to 50	100%	12
10 to 25	100%	24
0 to 10	50%	24

Hemodialysis decreases plasma concentrations 60% in a 6-hour period.

Oral:

Acute Treatment of Herpes Zoster: 800 mg every 4 hours orally, 5 times daily for 7 to 10 days

Genital Herpes:
Treatment of initial herpes: 200 mg every 4 hours, 5 times daily for 10 days
Chronic suppressive therapy for recurrent disease: 400 mg 2 times daily for up to 12 months, followed by re-evaluation.

Continued ...

Acyclovir

Immuno-suppression

Antimicrobials

Cardiovascular Agents

Antiosteo-porosis Agents

Antiplatelets

Diabetes Agents

Ulcer Treatment

Diuretics

Other Agents

Dosage	Treatment of Chickenpox:

Children (2 years of age and older): 20 mg/kg per dose orally 4 times daily (80 mg/kg per day) for 5 days. Children over 40 kg should receive the adult dose for chickenpox.

Adults and children over 40 kg: 800 mg 4 times daily for 5 days.

Dosing Modifications for Renal Impairment

		Adjusted Dosage Regimen	
Normal Dosage Regimen	Creatinine Clearance (mL/min)	Dose (mg)	Dosing Interval (h)
200 mg every 4 hours	>10	200	every 4 hours, 5x daily
	0 to 10	200	every 12 hours
400 mg every 12 hours	>10	400	every 12 hours
	0 to 10	200	every 12 hours
800 mg every 4 hours	>25	800	every 4 hours, 5x daily
	10 to 25	800	every 8 hours
	0 to 10	800	every 12 hours

Editors' Notes:

Very effective for prevention of herpes simplex. Many centers use high-dose acyclovir to prevent CMV infection, but studies conflict as to its effectiveness.

Immuno-suppression

Antimicrobials

Cardiovascular Agents

Antiosteo-porosis Agents

Antiplatelets

Diabetes Agents

Ulcer Treatment

Diuretics

Other Agents

Valacyclovir

Brand Name	Valtrex®
Company	GlaxoSmithKline
Class	Valacyclovir is the hydrochloride salt of L-valyl ester of acyclovir, synthetic nucleoside analogue with antiviral properties against herpesvirus
Mechanism of Action	• Inhibits the replication of herpesvirus cells by interfering with DNA polymerase, incorporation and termination of the growing viral DNA chain
Indication	• Treatment of herpes zoster (shingles) • Treatment of genital herpes
Contraindication	• Hypersensitivity to acyclovir or valacyclovir
Warning	• Thrombotic thrombocytopenic purpura/hemolytic uremic syndrome (TTP/HUS), which has resulted in death, has occurred in immunocompromised patients receiving acyclovir therapy
Precautions	• Recommended dosage, frequency, and length of treatment should not be exceeded • Precipitation of acyclovir crystals in renal tubules can occur, potentially leading to renal tubular damage • Dosage adjustments should be based on estimated creatinine clearance • 1% of patients have experienced encephalopathic changes • Resistant strains have developed after repeated exposure to acyclovir • Pregnancy Category B
Adverse Reactions	• Transient elevations in serum creatinine and BUN • Nausea and/or vomiting • Itching, rash, or hives • Elevation of transaminases • Encephalopathic changes, including lethargy, obtundation, tremors, confusion, hallucinations, agitation, seizures, and coma
Drug Interactions	• Cimetidine • Probenecid
Formulation	• 500 mg caplets in bottles containing 42 caplets • 1 gram caplets in bottles containing 21 caplets
Dosage	Valtrex caplets may be given without regard to meals Herpes zoster: The recommended dosage for the treatment of herpes zoster is 1 gram orally 3 times daily for 7 days. Genital herpes: Initial episodes: The recommended dosage for treatment of initial genital herpes is 1 gram twice daily for 3 days. Recurrent episodes: The recommended dosage for the treatment of recurrent genital herpes is 500 mg twice daily for 3 days. Suppressive therapy: The recommended dosage for chronic suppressive therapy of recurrent genital herpes is 1 gram once daily. In patients with a history of 9 or fewer recurrences a year, an alternative dose is 500 mg once daily.

Continued ...

Valacyclovir

Immuno-suppression

Antimicrobials

Cardiovascular Agents

Antiosteo-porosis Agents

Antiplatelets

Diabetes Agents

Ulcer Treatment

Diuretics

Other Agents

Dosage				

Patients with Acute or Chronic Renal Impairment
In patients with reduced renal function, reduction in dosage is recommended (see Table)

Dosing Adjustments for Patients with Renal Impairment

Indications	Normal Dosage Regimen (Creatinine Clearance >/=50)	Creatinine Clearance (mL/min) 30-49	10-29	≤10
Herpes zoster	1 gram every 8 hours	1gram every 12 hours	1 gram every 24 hours	500 mg every 24 hours
Genital herpes:				
Initial treatment	1 gram every 12 hours	no reduction	1 gram every 24 hours	500 mg every 24 hours
Recurrent episodes	500 mg every 12 hours	no reduction	500 mg every 24 hours	500 mg every 24 hours
Suppressive therapy	1 gram every 24 hours	no reduction	500 mg every 24 hours	500 mg every 24 hours
Suppressive therapy	500 mg every 24 hours	no reduction	500 mg every 48 hours	500 mg every 48 hours

Hemodialysis: During hemodialysis, the half-life of acyclovir after administration of valacyclovir is approximately 4 hours. About one third of acyclovir in the body is removed by dialysis during a 4-hour hemodialysis session. Patients requiring hemodialysis should receive the recommended dose of valacyclovir after hemodialysis.

Peritoneal Dialyis: There is no information specific to administration of VALTREX in patients receiving peritoneal dialysis. The effect of chronic ambulatory peritoneal dialysis (CAPD) and continuous arteriovenous hemofiltration/dialysis (CAVHD) on acyclovir pharmacokinetics has been studied. The removal of acyclovir after CAPD and CAVHD is less pronounced than with hemodialysis, and the pharmacokinetic parameters closely resemble those observed in patients with ESRD not receiving hemodialysis. Therefore, supplemental doses of VALTREX should not be required following CAPD or CAVHD.

Editors' Notes:

Some centers use valacyclovir to prevent CMV infection. Valacyclovir is more expensive than acyclovir but may improve patient compliance due to ease of administration. Bioavailability for valacyclovir is approximately 55%, whereas acyclovir is approximately 10% for 800 mg dose, 20% for 400 mg dose, and 30% for 200 mg dose.

Valacyclovir 1 G orally has approximately the same AUC as acyclovir 5 mg/kg IV. (JAC ©2001 47, 855-6.)

Ganciclovir

Immuno-suppression

Antimicrobials

Cardiovascular Agents

Antiosteo-porosis Agents

Antiplatelets

Diabetes Agents

Ulcer Treatment

Diuretics

Other Agents

Brand Name	Cytovene®
Company	Roche Laboratories
Class	Synthetic nucleoside analogue of 2'-deoxyguanine, with antiviral properties against herpesvirus including human cytomegalovirus
Mechanism of Action	• Inhibits replication by interfering with viral DNA synthesis • CMV retinitis in immunocompromised patients
Indication	• Prevention of CMV disease in transplant recipients at risk for CMV disease
Contraindication	• Hypersensitivity to ganciclovir or acyclovir • CMV retinitis in immunocompromised patients
Warnings	• Should not be administered if the absolute neutrophil count is <500 cells/mm³ or the platelet count is <25,000 cells/mm³ • Should be used with caution in patients with pre-existing cytopenias or a history of cytopenic reactions to other agents or procedures • May impair fertility • Because of the mutagenic potential of ganciclovir, women of childbearing age should use effective contraception • Pregnancy Category C
Special Precautions	• Exceeding recommended dosage and infusion rate could result in increased toxicity • Dosage adjustment is required in presence of renal impairment • Phlebitis and pain may occur at injection site
Adverse Reactions	• Granulocytopenia • Thrombocytopenia • Diarrhea • Nausea • Vomiting • Confusion • Decreased creatinine clearance • Headache • Sepsis • Anemia • Fever • Rash • Abnormal liver function values
Drug Interactions	• Zidovudine • Probenecid • Imipenem-cilastatin • Dapsone • Pentamidine • Flucytosine • Vincristine • Vinblastine • Adriamycin • Amphotericin B • TMP-SMX • Other nucleoside analogues

Continued ..

Ganciclovir

Formulation	• 10 mL vial containing 500 mg • 250 mg capsules in bottles of 180 capsules • 500 mg capsules in bottles of 180 capsules
Dosage	*Treatment of CMV retinitis* Induction - IV 5 mg/kg (over 1 hour) every 12 hours for 14 to 21 days Maintenance - IV 5 mg/kg once daily, 7 days per week or 6 mg/kg 5 days per week PO 1000 mg three times daily with food *Prevention of CMV in Transplant Recipients* IV - start with 5 mg/kg (over 1 hour) every 12 hours for 14 to 21 days followed by IV 5 mg/kg once daily, 7 days per week or 6 mg/kg 5 days per week PO 1000 mg three times daily with food Duration of therapy depends on degree of immunosuppression; in clinical trials, treatment was continued until postop day 100 to 120

Dosing Adjustments for Patients with Renal Dysfunction

Creatinine clearance (mL/min)	Induction dose (mg/kg)	Dosing interval (hrs)	Maintenance dose (mg/kg)	Dosing interval (hrs)
≥70	5.0	12	5.0	24
50 to 69	2.5	12	2.5	24
25 to 49	2.5	24	1.25	24
10 to 24	1.25	24	0.625	24
<10	1.25	3x/week following hemodialysis	0.625	3x/week following hemodialysis

Oral:
1000 mg po tid for up to 3 months

Dosing Adjustments for Patients with Renal Dysfunction

Creatinine clearance (mL/min)	Cytovene Capsule Dosages
≥70	1000 mg tid or 500 mg q 3 hr
50 to 69	1500 mg qd or 500 mg tid
25 to 49	1000 mg qd or 500 mg bid
10 to 24	500 mg qd
<10	500 mg 3x/week following hemodialysis

Editors' Notes:

Data suggest that ganciclovir can decrease CMV disease when given prophylactically during OKT3 or ATG treatment. Cytovene capsules were shown to reduce the incidence of CMV disease in a randomized, double-blind trial in liver transplant recipients when given orally for 3 months post-transplant. Single-center studies have also shown efficacy in renal transplant recipients. The cost-benefit of oral CMV prophylaxis remains to be established.

Immunosuppression

Antimicrobials

Cardiovascular Agents

Antiosteoporosis Agents

Antiplatelets

Diabetes Agents

Ulcer Treatment

Diuretics

Other Agents

Valganciclovir

Immuno-suppression

Antimicrobials

Cardiovascular Agents

Antiosteo-porosis Agents

Antiplatelets

Diabetes Agents

Ulcer Treatment

Diuretics

Other Agents

Brand Name	Valcyte®
Company	Roche Laboratories
Class	• L-valyl ester prodrug of ganciclovir, synthetic analogue of 2'-deoxyguanine, with antiviral properties against herpes including human cytomegalovirus
Mechanism of Action	• Inhibits replication by interfering with viral DNA synthesis
Indications	• Treatment of cytomegalovirus retinitis in patients with acquired immunodeficiency syndrome
Contraindications	• Hypersensitivity to valganciclovir or ganciclovir
Warnings	• Should not be administered if the absolute neutrophil count is less than 500 cells/microliter or platelet count less than 25,000 cells/microliter or the hemoglobin is less than 8 g/dl • Should be used in caution in patients with pre-existing cytopenias or who have received or who are receiving myelosuppressive drugs or irratdiation • May impair fertility • Because of the mutagenic and tertogenic potential of gangciclovir, women of childbearing age should use effective contraception-Pregnancy Category C
Special Precaution	• Exceeding recommended dosage could result in increased toxicity. Dosage adjustment is required in the presence of renal impairment.
Adverse Reactions	• Anemia • Nausea • Fever • Sepsis • Confusion	• Decreased creatinine clearance • Diarrhea • Vomiting • Rash	• Headache • Granulocytopenia • Thrombocytopenia • Abnormal liver disease
Drug Interactions	• Zidovudine • Imipenem-cilastatin • Pentamidine • Vincristine • Adriamycin	• TMP-SMX • Probenecid • Dapsone • Flucytosine • Vinblastine	• Amphotericin B • Other nucleoside analogues
Formulations	• 450 mg tablets in bottles of 60 tablets
Dosage	Crcl (mL/min) induction dose maintenance dose ≥ 60 900 mg twice daily 900 mg once daily 40-59 450 mg twice daily 450 mg once daily 25-39 450 mg once daily 450 mg every 2 days 10-24 450 mg every 2 days 450 mg twice weekly Hemodialysis reduces plasma concentration of ganciclovir by 50% following valganciclovir administration, patients receiving hemodialysis should not use valganciclovir because the daily dose is less than 450 mg.

Editors' Notes:

Many centers use valganciclovir to prevent CMV infection and disease. Valganciclovir 900 mg given with food has approximately the same AUC as ganciclovir 5 mg/kg intravenously. Valganciclovir 450 mg given with food has approximately the same AUC as ganciclovir 1000 mg given with food three times daily.

Valganciclovir compared to ganciclovir for prophylaxis of CMV following solid organ transplantation was equivalent in a randomized, controlled trial in donor positive/recipient negative transplant. Better pharmacokinetics and less frequent doses have abrogated the need for oral ganciclovir. Paya C et al. Am J Transplant 2004; 4(4):611-620.

Cytomegalovirus (CMV) Immune Globulin

Brand Name	**Cytogam®**
Company	MedImmune, Inc.
Class	• Immunoglobulin G preparation with antiviral properties
Mechanism of Action	• Contains a high concentration of IgG antibodies directed at CMV and, in individuals exposed to CMV, can elevate CMV-specific antibodies to levels sufficient to attenuate incidence of serious CMV disease
Indication	• Attenuation of primary CMV infection associated with renal, lung, liver, pancreas and heart transplantation recipients who are CMV-negative and receive an organ from a CMV-positive donor
Contraindications	• Hypersensitivity to human immunoglobulin preparations • Selective IgA deficiency
Warning	• During administration, patients' vital signs should be monitored closely and epinephrine should be available for treatment of an acute anaphylactic reaction
Special Precautions	• If hypotension or anaphylaxis occurs, treatment should be discontinued immediately • Does not contain a preservative and, accordingly, the vial should be entered only once after reconstitution • Pregnancy Category C
Adverse Reactions	• Flushing • Wheezing • Back pain • Muscle cramps • Chills • Nausea/vomiting • Fever
Drug Interactions	• May interfere with the immune response to live virus vaccines • Drug interactions have not been evaluated. However, it is recommended that immunoglobulin be administered separately from other agents
Formulation	• Single dose vial containing 2500 mg ± 250 mg or 1000 mg ± 200 mg of lyophilized immunoglobulin

Dosage

• Type of transplant

	Kidney	Liver, Pancreas, Lung, Heart
Within 72 hours of transplant	150 mg/kg	150 mg/kg
2 weeks post transplant	100 mg/kg	150 mg/kg
4 weeks post transplant	100 mg/kg	150 mg/kg
6 weeks post transplant	100 mg/kg	150 mg/kg
8 weeks post transplant	100 mg/kg	150 mg/kg
12 weeks post transplant	50 mg/kg	100 mg/kg
16 weeks post transplant	50 mg/kg	100 mg/kg

ABW is used to calculate dose

Initial dose—Administer intravenously at 15 mg Ig per kg body weight per hour. If no adverse reactions occur after 30 minutes, the rate may be increased to 30 mg Ig/kg/hr; if no adverse reactions occur after a subsequent 30 minutes, then the infusion may be increased to 60 mg Ig/kg/hr (volume not to exceed 75 mL/hour). DO NOT EXCEED THIS RATE OF ADMINISTRATION. The patient should be monitored closely during and after each rate change.

Editors' Notes:

Effective at preventing CMV disease, but more expensive than prophylactic ganciclovir. May also be useful as adjunctive therapy to ganciclovir in CMV disease.

CMV IVIg has been used in desensitization protocols to reduce donor specific antibodies.

Sidebar (vertical tabs): Immuno-suppression · Antimicrobials · Cardiovascular Agents · Antiosteo-porosis Agents · Antiplatelets · Diabetes Agents · Ulcer Treatment · Diuretics · Other Agents

Foscarnet Sodium

Immuno-suppression

Antimicrobials

Cardiovascular Agents

Antiosteo-porosis Agents

Antiplatelets

Diabetes Agents

Ulcer Treatment

Diuretics

Other Agents

Brand Name	**Foscavir®**
Company	Astra USA, Inc.
Class	• Organic analogue of inorganic pyrophosphate that inhibits replication of all known herpesviruses in vitro
Mechanism of Action	• Selective inhibition at the pyrophosphate binding site on virus-specific DNA polymerases and reverse transcriptases
Indication	• Treatment of CMV retinitis in patients with AIDS
Contraindication	• Hypersensitivity to foscarnet sodium
Warnings	• Renal impairment is the major toxicity of foscarnet and it occurs to some degree in most patients • Dose adjustments are necessary for patients with baseline renal dysfunction and in patients who experience changes in renal function during therapy • Changes in serum electrolytes may occur • Neurotoxicity and seizures have been reported in association with foscarnet therapy
Special Precautions	• Should be administered at recommended doses and infusion rates • Anemia may occur during therapy • Pregnancy Category C
Adverse Reactions	• Fever • Nausea • Anemia • Diarrhea • Abnormal renal function, including acute renal failure, decreased serum creatinine clearance, and increased serum creatinine • Vomiting • Headache • Seizure
Drug Interactions	• Pentamidine • Drugs that impair renal tubular secretions (aminoglycosides, amphotericin B) • Drugs known to affect serum calcium levels
Formulations	• 24 mg/mL in 250 mL bottle • 24 mg/mL in 500 mL bottle

Continued ...

Foscarnet Sodium

Immuno-suppression

Antimicrobials

Cardiovascular Agents

Antiosteo-porosis Agents

Antiplatelets

Diabetes Agents

Ulcer Treatment

Diuretics

Other Agents

Dosage

Rate of infusion must be carefully controlled with infusion pump

Induction—Normal Renal Function
- 60 mg/kg IV at a constant rate over a minimum of 1 hour every 8 hours for 2 to 3 weeks depending on clinical response

Maintenance—Normal Renal Function
- 90 mg/kg/d to 120 mg/kg/d IV infused over 2 hours

Dose Adjustments—Impaired Renal Function

Induction

CrCl (mL/min/kg)	Equivalent to 60 mg/kg q8h
≥1.6	60
1.5	57
1.4	53
1.3	49
1.2	46
1.1	42
1.0	39
0.9	35
0.8	32
0.7	28
0.6	25
0.5	21
0.4	18

Maintenance

CrCl (mL/min/kg)	Equivalent to 90 mg/kg/d	Equivalent to 120 mg/kg/d
1.4	90	120
1.2 to 1.4	78	104
1.0 to 1.2	75	100
0.8 to 1.0	71	94
0.6 to 0.8	63	84
0.4 to 0.6	57	76

Immuno-suppression

Antimicrobials

Cardiovascular Agents

Antiosteo-porosis Agents

Antiplatelets

Diabetes Agents

Ulcer Treatment

Diuretics

Other Agents

Cidofovir

Brand Name	Vistide®
Company	Pharmacia & Upjohn
Class	• Cytosine nucleoside analog
Mechanism of Action	• Selectively inhibits viral (CMV/herpes simplex types1 and 2) DNA polymerase and reduces in the rate of viral DNA synthesis
Indications	• Treatment of CMV retinitis in patients with AIDS
Application in Transplantation	• Adjunct therapy for treatment of BK nephropathy
Contraindications	• Hypersensitivity to cidofovir • Serum creatinine greater than 1.5 mg/dl, a calculated CrCl of 55 mL/min or less, or a urine protein of 100 mg/dl (equivalent to 2+ proteinuria) • Concomitant nephrotoxic agents • Direct intraocular injection of cidofovir
Warnings	• Renal impairment (cases of acute renal failure resulting dialysis) is the major toxicity of cidofovir • Neutropenia has been observed • Cidofovir is carcinogenic, teratogenic and causes hypospermia in animal studies
Special Precautions	• Nephrotoxicity; intravenous pre-hydration and post-hydration with normal saline and probenecid 2 grams orally 3 hours prior to cidofovir infusion, then 1 gram at 2 hours and again 8 hours after the 1-hour infusion (total of 4 grams probenecid) should be used • Monitor for decreased intraocular pressure • Monitor appropriately for neutrophil counts • Pregnancy Category C
Adverse Reactions	• Dose-dependent nephrotoxicity due to proximal tubular dysfunction (dose-limiting) • Elevated serum creatinine • Nausea/vomiting/diarrhea • Neutropenia • Headache/asthenia • Peripheral neuropathy • Iritis/uveitis • Intraocular pressure changes • Hearing loss • Metabolic acidosis (<1%) • Nephrogenic diabetes insipidus (case report)
Drug Interactions	*Drugs with Synergistic Nephrotoxicity* • Amikacin • Tobramycin • Foscarnet • Pentamidine (IV) • Gentamicin *Drugs that Increase Cidofovir Levels* • Tenofovir disoproxil fumarate
Formulation	• 5-mL vials containing 375 mg

Continued ...

Cidofovir

Immuno-suppression

Antimicrobials

Cardiovascular Agents

Antiosteo-porosis Agents

Antiplatelets

Diabetes Agents

Ulcer Treatment

Diuretics

Other Agents

Dosage

CMV Retinitis in AIDS Patients
- Induction phase—5 mg/kg IV once weekly for two weeks
- Maintenance phase—5 mg/kg IV once every other week until retinitis progression or therapy-limiting toxicity

Mucocutaneous Herpes Simplex Infection Resistant to Acyclovir
- 5 mg/kg IV weekly (limited data)

BK Nephropathy
- 0.25-1 mg/kg IV every other week (probenecid should not be given) (limited data)

Editors' Notes:

The major use for cidofovir in clinical transplantation is in the treatment of BK virus nephropathy. The doses used are much lower than for CMV infection and usually 0.25-1 mg/kg weekly.

Lamivudine

Brand Name	**Epivir-HBV®**
Company	GlaxoSmithKline
Class	• Synthetic nucleoside analogue with activities against hepatitis B virus (HBV) and HIV
Mechanism of Action	• Inhibits the replication of HBV/HIV by interfering with viral DNA polymerase and terminating viral DNA chain synthesis
Application in Transplantation	• Suppresses HBV and improves liver function test as a post liver transplantation treatment option in patients with chronic HBV infections
Contraindications	• Hypersensitivity to lamivudine and any of its components
Warnings	• Lactic acidosis and severe hepatomegaly with steatosis, which has resulted in death, has been observed • Exacerbation of hepatitis has occurred after discontinuation of therapy with lamivudine • Pancreatitis has been reported
Special Precautions	• Epivir® (for treatment of HIV infection) contains a higher dose of lamivudine than Epivir-HBV® • Female gender, obesity, liver diseases and prolonged nucleoside exposure can increase risk for lactic acidosis and severe hepatomegaly with steatosis • Dose reduction is recommended in patients with renal impairment • Hepatitis progression can occur in patients with HBV mutation • Pregnancy Category C
Adverse Reactions	• Lactic acidosis/severe hepatomegaly with steatosis • Elevated serum creatinine • Nausea/vomiting/diarrhea/abdominal pain • Pancreatitis • Neutropenia • Anemia • Arthralgia/myalgia/musculoskeletal pain • Cough • Malaise/fatigue/dizziness/headache/insomnia • Neuropathy • Body fat redistribution/accumulation • Hearing loss • Fevers/chills (10%) • Rash/pruritus (10%)
Drug Interactions	*Drugs causing hepatic decompensation* • Interferon alfa • Ribavirin *Drugs that Increase lamivudine exposure* • Trimethoprim • Sulfamethoxazole • Zalcitabine
Formulation	• 100 mg tablets in bottles of 60 • 5 mg/mL oral solution in plastic bottles of 240 mL

Continued ...

Lamivudine

Immuno-suppression

Antimicrobials

Cardiovascular Agents

Antiosteo-porosis Agents

Antiplatelets

Diabetes Agents

Ulcer Treatment

Diuretics

Other Agents

Dosage	*Chronic Hepatitis B Infection* • 100 mg orally once daily (optimal duration of therapy unknown)

Dosage Adjustment for Patients with Renal Impairment

Creatinine Clearance (mL/min)	Dosage
≥ 50	100 mg once daily
30–49	First dose 100 mg then 50 mg once daily
15–29	First dose 100 mg then 25 mg once daily
5–14	First dose 35 mg then 15 mg once daily
< 5	First dose 35 mg then 10 mg once daily

Editors' Notes:

Effective in suppressing HBV replication, but a significant number of patients develop breakthrough infection with HBV mutant viruses.

Adefovir Dipivoxil

Immuno-suppression

Antimicrobials

Cardiovascular Agents

Antiosteo-porosis Agents

Antiplatelets

Diabetes Agents

Ulcer Treatment

Diuretics

Other Agents

Brand Name	Hepsera®
Company	Gilead
Class	• Acyclic nucleotide reverse transcriptase inhibitor with activities against hepatitis B virus (HBV) and HIV
Mechanism of Action	• Inhibits the replication of HBV/HIV by interfering with viral DNA polymerase and terminating viral DNA chain synthesis
Application in Transplantation	• Suppresses HBV and improves liver function test as a post liver transplantation treatment option in patients with chronic HBV infections
Contraindications	• Hypersensitivity to adefovir and any of its components
Warnings	• Lactic acidosis and severe hepatomegaly with steatosis, which has resulted in death, has been observed • Severe exacerbation of hepatitis has occurred after discontinuation of anti-HBV therapy with adefovir • Nephrotoxicity with chronic administration in patients at risk of or have underlying renal dysfunction
Special Precautions	• Risk of emergence of HIV resistance in co-infected patients • Dose adjustment is recommended in patients with renal impairment • Pregnancy Category C
Adverse Reactions	• Lactic acidosis/severe hepatomegaly with steatosis • Pharyngitis/sinusitis • Nausea/vomiting/diarrhea/dyspepsia • Increased liver function test • Hepatic failure • Cough • Asthenia/headache • Nephrotoxicity • Rash/pruritus (10%)
Drug Interactions	• Tenofovir disoproxil fumarate
Formulation	• 10 mg tablets in bottles of 30
Dosage	*Chronic Hepatitis B Infection* • 10 mg orally once daily (optimal duration of therapy unknown) without regard to food *Dosage Adjustment for Patients with Renal Impairment*

Creatinine Clearance (mL/min)	Dosage
≥ 50	10 mg once daily
20 – 49	10 mg every 2 days
10 -29	10 mg every 3 days
< 10	No dosing recommendation
Hemodialysis	10 mg every 7 days following dialysis

Editors' Notes:

Adefovir has been shown to be superior to lamivudine in preventing HBV mutations.

Oseltamivir Phosphate

Brand Name	Tamiflu®
Company	Roche
Class	• Neuraminidase inhibitor
Mechanism of Action	• Inhibition of influenza virus neuraminidase and causing alterations of virus particle aggregation and release
Indications	• Treatment of influenza infection in patients 1 year and older who have been symptomatic for no more than 2 days
Contraindications	• Hypersensitivity to oseltamivir phosphate or any of its components
Special Precautions	• Efficacy of treatment beginning 40 hours after onset of symptoms has not been established • Efficacy in patients with chronic cardiac disease, chronic respiratory disease, or immunosuppression; has not been established • Safety has not been established in patients with hepatic impairment • Pregnancy Category C
Adverse Reactions	• Nausea
	• Vomiting • Dizziness • Fatigue • Headache • Insomnia
Drug Interactions	• No clinically significant drug interactions
Formulation	• 75 mg gelatin capsules • Suspension
Dosage	• 75 mg orally twice daily for 5 days

Immuno-suppression

Antimicrobials

Cardiovascular Agents

Antiosteo-porosis Agents

Antiplatelets

Diabetes Agents

Ulcer Treatment

Diuretics

Other Agents

Antibiotics Overview

Immuno-suppression

Antimicrobials

Cardiovascular Agents

Antiosteo-porosis Agents

Antiplatelets

Diabetes Agents

Ulcer Treatment

Diuretics

Other Agents

Agent	Dosage
TMP-SMX	*IV Infusion* • *Pneumocystis carinii* pneumonia—15 mg/kg/d to 20 mg/kg/d given in 3 or 4 equally divided doses every 6 to 8 hours for up to 14 days • UTI/Shigellosis—8 mg/kg/d to 10 mg/kg/d given in 2 or 4 equally divided doses every 6, 8, or 12 hours for up to 14 days for severe UTIs and for 5 days for shigellosis *Tablets* • *Pneumocystis carinii* pneumonia—15 mg/kg/d to 20 mg/kg/d TMP and 75 mg/kg/d to 100 mg/kg/d SMX in equally divided doses every 6 hours for 14 to 21 days • Chronic bronchitis—2 tablets every 12 hours for 14 days • Travelers' diarrhea—2 tablets every 12 hours for 5 days • UTI/Shigellosis—2 tablets every 12 hours for 10 to 14 days for UTIs and for 5 days for shigellosis • Otitis media (children)—8 mg/kg/d TMP and 40 mg/kg/d SMX given in 2 equally divided doses every 12 hours for 10 days
Penicillins with Beta-Lactamase Inhibitor	*Amoxicillin/Clavulanate potassium* • One 250 mg tablet every 8 hours • One 500 mg tablet every 8 hours for serious infection • One 875 mg tablet every 12 hours for serious infection *Ticarcillin/Clavulanate potassium* • 3.1 g vial every 4 to 6 hours for systemic and urinary tract infections • 200 mg/kg/d in divided doses every 6 hours for moderate gynecologic infections • 300 mg/kg/d in divided doses every 4 hours for severe gynecologic infections *Pipercillin/Tazobactam sodium* *Nosocomial Pneumonia* • 4.5 grams intravenously every 6 hours in combination with an aminoglycoside *Other Infections* • 3.375 grams intravenously every 6 hours *Ampicillin/Sulbactam sodium* • 1.5-3 grams intravenously every 6 hours
Quinolones	*Levofloxacin* • 250 mg daily for complicated urinary tract infection • 500 mg daily for acute maxillary sinusitis and community-acquired pneumonia or uncomplicated skin infections *Ciprofloxacin Tablets* • 250 mg every 12 hours for urinary tract infection; patients with complicated infections may require 500 mg every 12 hours • A single 250 mg dose for urethral and cervical gonococcal infections • 500 mg every 12 hours for lower respiratory infection, skin and skin structure infection, infectious diarrhea, and bone and joint infection; serious infection may require 750 mg every 12 hours • Duration of therapy is usually 7 to 14 days; bone and joint infections may require therapy for 4 to 6 weeks; infectious diarrhea may be treated for 5 to 7 days

Continued ...

Immuno-suppression

Antimicrobials

Cardiovascular Agents

Antiosteo-porosis Agents

Antiplatelets

Diabetes Agents

Ulcer Treatment

Diuretics

Other Agents

Antibiotics Overview

Agent	Dosage
Quinolones (continued)	*Ciprofloxacin* *Moxifloxacin Hydrochloride* • 400 mg orally or IV daily *Gatifloxacin* • 400 mg orally or IV daily *Levofloxacin* • 250 mg-750 mg orally or IV daily *IV Infusion* • 200 mg every 12 hours for mild to moderate urinary tract infections; 400 mg every 12 hours for severe urinary tract infections • 400 mg every 12 hours for lower respiratory infection, skin and skin structure infection, and bone and joint infection • Duration of therapy is usually 7 to 14 days; bone and joint infections may require therapy for 4 to 6 weeks
Cephalosporins	*Cefazolin* • Treatment • moderate/severe 500 mg to 1 g every 6 to 8 hours • mild 250 mg to 500 mg every 8 hours • UTIs 1 g every 12 hours • Pneumonia 500 mg every 12 hours • Severe/life-threatening 1 g to 1.5 g every 6 hours • Prophylaxis 1 g IV or IM administered once 30 to 60 minutes prior to surgery; 500 mg to 1 g IV or IM may be given intraoperatively during lengthy procedures; 500 mg to 1 g IV or IM every 6 to 8 hours for 24 hours postop *Cefotetan* • Treatment 1 g or 2 g IV or IM every 12 hours for 5 to 10 days • Prophylaxis 1 g or 2 g IV administered once 30 to 60 minutes prior to surgery *Ceftriaxone* • Treatment 1-2 gm IV once or twice daily for 4 to 14 days • Prophylaxis 1 g IV 30 to 120 minutes prior to surgery *Cefepime* • Treatment 1 to 2 g IV q 12 hours for 10 days

Continued ...

Antibiotics Overview

Agent	Dosage
Aminoglycosides	*Amikacin—Patients with Normal Renal Function* • 15 mg/kg/d IM or IV divided into 2 or 3 equal doses administered at equally divided intervals • Usual duration of treatment is 7 to 10 days and it is desirable to limit the duration to the shortest time possible *Tobramycin/Gentamicin—Patients with Normal Renal Function* • 3 mg/kg/d IM or IV divided into 3 equal doses every 8 hours • Patients with life-threatening infections may be given up to 5 mg/kg/d in 3 or 4 equal doses, with a reduction to 3 mg/kg/d as soon as clinically indicated • Usual duration of treatment is 7 to 10 days and it is desirable to limit the duration to the shortest time possible
Linezolid	• 600 mg every 12 hours orally or intravenously
Tigecycline	• Initial dose 100 mg intravenously, followed by 50 mg intravenously every 12 hours
Daptomycin	• 4 mg/kg IV every 24 hours

Immuno-suppression

Antimicrobials

Cardiovascular Agents

Antiosteo-porosis Agents

Antiplatelets

Diabetes Agents

Ulcer Treatment

Diuretics

Other Agents

Immuno-suppression

Antimicrobials

Cardiovascular Agents

Antiosteo-porosis Agents

Antiplatelets

Diabetes Agents

Ulcer Treatment

Diuretics

Other Agents

Trimethoprim and Sulfamethoxazole (TMP-SMX)

Brand Name	Septra®	Bactrim™
Company	Monach	Roche Laboratories
Class	• Synthetic antibacterial combinations with activity against *Escherichia coli*, *Proteus* species, *Morganella morganii*, *Proteus mirabilis*, *Klebsiella* species, *Enterobacter* species, *Hemophilus influenzae*, *Streptococcus pneumoniae*, *Shigella flexneri*, and *Shigella sonnei*	
Mechanism of Action	• TMP-SMX blocks two consecutive steps in the biosynthesis of nucleic acids and proteins essential to many bacteria	
Indications	*IV Infusion* • Prophylaxis and treatment of *Pneumocystis carinii* pneumonia • Shigellosis • Severe or complicated urinary tract infections *Tablets* • Prophylaxis and treatment of *Pneumocystis carinii* pneumonia • Acute exacerbation of chronic bronchitis • Urinary tract infections • Travelers' diarrhea in adults • Shigellosis • Acute otitis media	
Contraindications	• Hypersensitivity to TMP or SMX • Megaloblastic anemia due to folate deficiency • Pregnancy at term and nursing • Infants <2 months of age	

Warnings	• Deaths due to following severe reactions have occurred after treatment with SMX
	  • Stevens-Johnson syndrome
	  • Toxic epidermal necrolysis
	  • Fulminant hepatic necrosis
	  • Agranulocytosis
	  • Aplastic anemia
	  • Other blood dyscrasias
	  • Hypersensitivity of the respiratory tract
	• Should not be used for the treatment of streptococcal pharyngitis	

Special Precautions	• Should be given with caution to patients with renal or hepatic impairment, folate deficiency, and severe allergy or asthma • Pregnancy Category C	
Adverse Reactions	• Gastrointestinal disturbances (nausea, vomiting, anorexia) • Allergic skin reactions (rash, urticaria) • Agranulocytosis, aplastic anemia, and other blood dyscrasias • Stevens-Johnson syndrome, toxic epidermal necrolysis, and other allergic reactions • Hepatitis, including cholestatic jaundice and hepatic necrosis • Renal failure and interstitial nephritis • Aseptic meningitis, convulsions, peripheral neuritis, ataxia, vertigo, tinnitus, and headache • Hallucinations, depression, apathy, and nervousness	

Continued ..

Trimethoprim and Sulfamethoxazole (TMP-SMX)

Immuno-suppression

Antimicrobials

Cardiovascular Agents

Antiosteo-porosis Agents

Antiplatelets

Diabetes Agents

Ulcer Treatment

Diuretics

Other Agents

Drug Interactions	• Thiazides • Warfarin • Phenytoin • Methotrexate • Serum methotrexate assays • Jaffe alkaline picrate reaction assay	
Formulations	*IV Infusion* • 5 mL vial containing 80 mg TMP (16 mg/mL) and 400 mg (80 mg/mL) SMX • 10 mL vial containing 160 mg (16 mg/mL) TMP and 800 mg (80 mg/mL) SMX • 20 mL vial containing 320 mg (16 mg/mL) TMP and 1600 mg (80 mg/mL) SMX *Oral Suspension* • 40 mL TMP and 200 mg SMX per 5 mL suspension *Tablets* • Tablets containing 80 mg TMP and 400 mg SMX supplied in bottles of 100 • Double-strength tablets containing 160 mg TMP and 800 mg SMX supplied in bottles of 100	
Dosage	*IV Infusion* • *Pneumocystis carinii* pneumonia	15 mg/kg/d to 20 mg/kg/d given in 3 or 4 equally divided doses every 6 to 8 hours for up to 14 days
	• UTI/Shigellosis	8 mg/kg/d to 10 mg/kg/d given in 2 or 4 equally divided doses every 6, 8, or 12 hours for up to 14 days for severe UTIs and for 5 days for shigellosis
	Tablets • *Pneumocystis carinii* pneumonia	15 mg/kg/d to 20 mg/kg/d TMP and 75 mg/kg/d to 100 mg/kg/d SMX in equally divided doses every 6 hours for 14 to 21 days
	• Chronic bronchitis	2 tablets every 12 hours for 14 days
	• Travelers' diarrhea	2 tablets every 12 hours for 5 days
	• UTI/Shigellosis	2 tablets every 12 hours for 10 to 14 days for UTIs and for 5 days for shigellosis
	• Otitis media (children)	8 mg/kg/d TMP and 40 mg/kg/d SMX given in 2 equally divided doses every 12 hours for 10 days

Editors' Notes:

Best randomized trials:

Fox BC, Sollinger HW, Belzer FO, Maki DG. A prospective, randomized, double-blind study of trimethoprim-sulfamethoxazole for prophylaxis of infection in renal transplantation: clinical efficacy, absorption of trimethoprim-sulfamethoxazole, effects on the microflora, and the cost-benefit of prophylaxis. Am J Med 1990; 89:225-274.

Maki DG, Fox BC, Gould JR, Sollinger HW, Belzer FO. A prospective, randomized, double-blind study of trimethoprim-sulfamethoxazole for prophylaxis of infection in renal transplantation: side effects of trimethoprim-sulfamethoxazole, interaction with cyclosporine-A. J Lab Clin Med 1992; 119(1):11-24.

Immuno-suppression

Antimicrobials

Cardiovascular Agents

Antiosteo-porosis Agents

Antiplatelets

Diabetes Agents

Ulcer Treatment

Diuretics

Other Agents

Penicillins with Beta-Lactamase Inhibitor

Brand Name	• Augmentin® (amoxicillin/clavulanate potassium) • Timentin® (ticarcillin/clavulanate potassium)
Company	• GlaxoSmithKline
Class	• Synthetic antibacterial combinations containing a penicillin (amoxicillin/ticarcillin) and a beta-lactamase inhibitor • Both have activity against *Staphylococcus aureus, Staphylococcus epidermidis, Staphylococcus saprophyticus, Streptococcus faecalis, Streptococcus pneumoniae, Streptococcus pyogenes, Streptococcus viridans, Clostridium* species, *Peptococcus* species, *Hemophilus influenzae, Moraxella catarrhalis, Escherichia coli, Klebsiella* species, *Enterobacter* species, *Proteus mirabilis, Proteus vulgaris, Neisseria gonorrhoeae, Legionella* species, and *Bacteroides species* • Ticarcillin/Clavulanate potassium also has activity against *Pseudomonas* species, *Providencia rettgeri, Providencia stuartii, Morganella morganii, Acinetobacter* species, *Serratia* species, *Neisseria meningitidis, Salmonella* species, *Citrobacter* species, *Streptococcus bovis*, and *Streptococcus agalactiae*
Mechanism of Action	• Clavulanic acid protects penicillins from degradation by inactivating a wide range of beta-lactamase enzymes
Indications	*Amoxicillin/Clavulanate potassium* • Lower respiratory infection • Otitis media • Sinusitis • Skin and skin structure infection • Urinary tract infection *Ticarcillin/Clavulanate potassium* • Septicemia • Lower respiratory infection • Bone and joint infection • Skin and skin structure infection • Urinary tract infection • Gynecologic infection • Intra-abdominal infection
Contraindication	• History of allergic reactions to penicillins
Warnings	• Serious and occasionally fatal anaphylactoid reactions have been reported in patients on penicillin therapy • Serious anaphylactoid reactions require immediate emergency treatment with epinephrine • Pseudomembranous colitis has been reported
Special Precautions	• Superinfections with mycotic or bacterial pathogens may occur • Pregnancy Category B

Continued …

Penicillins with Beta-Lactamase Inhibitor

Immuno-suppression

Antimicrobials

Cardiovascular Agents

Antiosteo-porosis Agents

Antiplatelets

Diabetes Agents

Ulcer Treatment

Diuretics

Other Agents

Adverse Reactions	• Skin rash, pruritus, urticaria, arthralgia, myalgia, drug fever, chills, chest discomfort, anaphylactic reactions • Headache, giddiness, neuromuscular hyperirritability, convulsive seizures • Taste and smell disturbances, stomatitis, flatulence, nausea, vomiting, diarrhea, epigastric pain, pseudomembranous colitis • Thrombocytopenia, leukopenia, neutropenia, reduction of hematocrit and hemoglobin • Abnormal hepatic and renal function tests • Pain, burning, swelling, and induration at injection site
Drug Interactions	• Aminoglycosides • Probenecid • False-positive protein reactions • False-positive Coombs' test
Formulations	*Amoxicillin/Clavulanate potassium* • 250 mg tablets (250 mg amoxicillin/125 mg clavulanic acid) • 500 mg tablets (250 mg amoxicillin/125 mg clavulanic acid) • 875 mg tablets (250 mg amoxicillin/125 mg clavulanic acid) • Extended-release tablets (1000 mg amoxicillin and 62.5 mg clavulanic acid) • 125 mg chewable tablet (125 mg amoxicillin and 31.25 mg clavulanic acid) • 200 mg chewable tablet (200 mg amoxicillin and 28.5 mg clavulanic acid) • 250 mg chewable tablet (250 mg amoxicillin and 62.5 mg clavulanic acid) • 400 mg chewable tablet (400 mg amoxicillin and 57 mg clavulanic acid) • 125 mg/5 mL for oral suspension (125 mg amoxicillin and 31.25 mg clavulanic acid) • 200 mg/5 mL for oral suspension (200 mg amoxicillin and 28.5 mg clavulanic acid) • 250 mg/5 mL for oral suspension (250 mg amoxicillin and 62.5 mg clavulanic acid) • 400 mg/5 mL for oral suspension (400 mg amoxicillin and 57 mg clavulanic acid) • 600 mg/5 mL for oral suspension (600 mg amoxicillin and 42.9 mg clavulanic acid) *Ticarcillin/Clavulanate potassium* • 3.1 gram vials, Piggyback bottles, ADD-Vantage vials • 31 gram pharmacy bulk packages • 3.1 gram premixed bags
Dosage	*Amoxicillin/Clavulanate potassium* • One 250 mg tablet every 8 hours • One 500 mg tablet every 8 hours for serious infection • One 875 mg tablet every 12 hours for serious infection *Ticarcillin/Clavulanate potassium* • 3.1 g vial every 4 to 6 hours for systemic and urinary tract infections • 200 mg/kg/d in divided doses every 6 hours for moderate gynecologic infections • 300 mg/kg/d in divided doses every 4 hours for severe gynecologic infections

Editors' Notes:

Augmentin and Timentin are frequently used for diabetic foot infections and urinary tract infections.

Penicillins with Beta-Lactamase Inhibitor

Immuno-suppression

Antimicrobials

Cardiovascular Agents

Antiosteo-porosis Agents

Antiplatelets

Diabetes Agents

Ulcer Treatment

Diuretics

Other Agents

Brand Name	Unasyn® (ampicillin/sulbactam sodium)
Company	Pfizer
Class	• Antibacterial combination containing a semisynthetic penicillin (ampicillin) and a beat-lactamase inhibitor • Active against *Staphylococcus aureus, Staphylococcus epidermidis, Streptococcus pneumoniae, Streptococcus pyogenes, Streptococcus viridans, Enterococcus faecali,, Escherichia coli, Haemophilus influenzae, Klebsiella species, Bacteroides species, Moraxella catarrhalis, Neisseria gonorrhoeae, Proteus mirabilis, Proteus vulgaris, Salmonella species, Clostridium species,* and *Listeria monocytogenes*
Mechanism of Action	• Ampicillin exerts bactericidal activity via inhibition of bacterial cell wall synthesis and sulbactam sodium protects ampicillin from degradation by inactiviating a wide range of bacterial beta-lactamases
Indications	• Gynecological infection • Skin and subcutaneous tissue infection • Intra-abdominal infection
Contraindications	• History of allergic reactions to any of the penicillins or any of its components
Warnings	• Serious and occasionally fatal anaphylactic reactions have been reported in patients receiving penicillin therapy • Serious anaphylaxis reactions require immediate emergency treatment with epinephrine • Pseudomembranous colitis has been reported
Special Precautions	• Neuromuscular excitability or seizure activity (especially at high doses or in patients with renal impairment) • Pregnancy Category B
Adverse Reactions	• Thrombophlebitis • Injection site discomfort • Rash • Diarrhea • Hypersensitivity reactions • Hearing loss (in high doses) • Fungal superinfection • Headache/malaise/fatigue
Drug Interactions	• Aminoglycosides • Probenecid • Oral contraceptives
Formulation	• 1.5 g vials and ADD-Vantage® vials • 3 g vials and ADD-Vantage® vials
Dosage	• 1.5-3 g IV/IM every 6 hr

Penicillins with Beta-Lactamase Inhibitor

Immuno-suppression

Antimicrobials

Cardiovascular Agents

Antiosteo-porosis Agents

Antiplatelets

Diabetes Agents

Ulcer Treatment

Diuretics

Other Agents

Brand Name	Zosyn® (pipercillin/tazobactam sodium)
Company	Wyeth
Class	• Antibacterial combination containing a semisynthetic penicillin (pipercillin) and a beat-lactamase inhibitor • Active against *Staphylococcus aureus, Staphylococcus epidermidis, Streptococcus agalactiae, Streptococcus pneumoniae, Streptococcus pyogenes, Streptococcus viridans, Enterococcus faecali, Acinetobacter baumanii, Escherichia coli, Haemophilus influenzae, Klebsiella pneumoniae, Pseudomonas aeruginosa, Bacteroides species, Citrobacter koseri, Moraxella catarrhalis, Morganella morganii, Neisseria gonorrhoeae, Proteus mirabilis, Proteus vulgaris, Serratia marcescens, Providencia stuartii, Providencia rettgeri, Salmonella enterica, Clostridium perfringens,* and *Prevotella melaninogenica*
Mechanism of Action	• Piperacillin exerts bactericidal activity via inhibition of bacterial cell wall synthesis and tazobactam sodium protects pipercillin from degradation by inactivating a wide range of bacterial beta-lactamases
Indications	• Intra-abdominal infection • Bone and joint infection • Peritonitis • Pelvic inflammatory disease • Skin and skin structure infections • Postpartum endometritis or pelvic inflammatory disease • Community-acquired pneumonia • Nosocomial pneumonia
Contraindications	• History of allergic reactions to any of the penicillins or any of its components
Warnings	• Serious and occasionally fatal anaphylactic reactions have been reported in patients receiving penicillin therapy • Serious anaphylaxis reactions require immediate emergency treatment with epinephrine • Pseudomembranous colitis has been reported
Special Precautions	• Neuromuscular excitability or seizure activity (especially at high doses or in patients with renal impairment) • 2.35 mEq of sodium per gram of pipercillin and tazobactam • Pregnancy Category B
Adverse Reactions	• Rash/pruritus • Diarrhea • Hemolytic anemia • Anemia/pancytopenia/leukocytopenia/ thrombocytopenia • Thrombocytosis • Hepatitis and cholestatic jaundice • Hypersensitivity reactions • Prolonged muscle relaxation/myalgia/arthralgia • Headache • Insomnia • Interstitial nephritis/renal failure • Drug fever

Continued ...

Immuno-suppression

Antimicrobials

Cardiovascular Agents

Antiosteo-porosis Agents

Antiplatelets

Diabetes Agents

Ulcer Treatment

Diuretics

Other Agents

Penicillins with Beta-Lactamase Inhibitor

Drug Interactions	• Aminoglycosides • Probenecid • Methotrexate • False-positive aspergillus galactomannan detection
Formulation	• 2.25 g vials and ADD-Vantage® vials • 3.375 g vials and ADD-Vantage® vials • 4.5 g vials and ADD-Vantage® vials
Dosage	*Nosococomial Pneumonia* • 4.5 g intravenously over 30 minutes every 6 hours in combination with an aminoglycoside *Other Infections* • 3.375 g intravenously over 30 minutes every 6 hours

Quinolones

Brand Name	Avelox® (moxifloxacin hydrochloride)	Levoquin® (levofloxacin)	Tequin®
Company	Schering	Ortho-McNeil	Bristol-Myers Squibb
Class	• Synthetic broad spectrum antibacterial agents • Active against *Staphylococcus aureus, Staphylococcus epidermidis, Streptococcus pneumoniae, Streptococcus pyogenes, Streptococcus viridans, Enterococcus faecali,, Escherichia coli, Haemophilus influenzae, Klebsiella species, Bacteroides species, Moraxella catarrhalis, Neisseria gonorrhoeae, Legionella pneumophila, Mycoplasma pneumoniae Proteus mirabilis, Proteus vulgaris, Salmonella species, Clostridium species,* and *Listeria monocytogenes*		
Mechanism of Action	• Inhibits bacterial DNA topoisomerases required for bacterial DNA replication, transcription, repair, and recombination		
Indications	*Moxifloxacin hydrochloride* • Community acquired pneumonia • Sinusitis • Acute bacterial exacerbation of chronic bronchitis • Uncomplicated and complicated skin and skin structure infection • Complicated intra-abdominal infection *Levofloxacin* • Community acquired pneumonia • Nosocomial pneumonia • Acute bacterial sinusitis • Acute bacterial exacerbation of chronic bronchitis • Uncomplicated and complicated skin and skin structure infection • Chronic bacterial prostatitis • Uncomplicated and complicated urinary tract infections • Acute pyelonephritis • Inhalational anthrax *Gatifloxacin* • Community acquired pneumonia • Sinusitis • Acute bacterial exacerbation of chronic bronchitis • Uncomplicated skin and skin structure infection • Urinary tract infection • Pyelonephritis • Uncomplicated urethral and cervical gonorrhea		
Contraindications	• History of allergic reactions to any of the quinolones or any of the components		
Warnings	• Safety and efficacy has not been established in children, adolescents, pregnant women, and lactating women • Potential for QT interval prolongation • Convulsions have been reported in patients receiving quinolones • Serious hypersensitivity have occurred • Pseudomembranous colitis has been reported • Peripheral neuropathy has been reported • Ruptures of the tendons that required surgical repair or resulted in prolonged disability have been reported		

Continued ...

Quinolones

Immuno-suppression

Antimicrobials

Cardiovascular Agents

Antiosteo-porosis Agents

Antiplatelets

Diabetes Agents

Ulcer Treatment

Diuretics

Other Agents

Special Precautions	• Avoid using in patients at risk for QT interval prolongation • Pregnancy Category C
Adverse Reactions	• Palpitation • Prolonged QT interval • Nausea/diarrhea/abdominal pain • Hyperglycemia/hypoglycemia • Hypersensitivity reactions • Elevated liver function tests • Rupture of tendon • Abnormal vision • Fungal superinfection • Headache/insomnia/asthenia/dizziness • Seizure
Drug Interactions	*Drugs that can increase QT interval* • Amiodarone • Amitriptyline/desipramine • Chlorpromazine • Disopyramide • Dofetilide • Erythromycin • Flecainide • Lidocaine • Quinidine *Other drug interactions* • Hypoglycemics • Antacids containing magnesium, aluminum, or calcium • Mineral supplements (iron, calcium) • Oral contraceptives • Warfarin
Formulation	*Moxifloxacin Hydrochloride* • 400 mg tablet • 400 mg in 250-mL bag *Gatifloxacin* • 200 mg tablet • 400 mg tablet • 400 mg/40 mL single dose vial • 200 mg ready-to-use flexible bag • 400 mg ready-to-use flexible bag *Levofloxacin* • 250 mg tablet • 500 mg tablet • 750 mg tablet • 25 mg/mL oral solution in 480-mL bottle • 500 mg/20 mL single dose vial • 750 mg/30 mL single dose vial • 250 mg ready-to-use flexible bag • 500 mg ready-to-use flexible bag • 750 mg ready-to-use flexible bag
Dosage	*Moxifloxacin Hydrochloride* • 400 mg orally or IV *Gatifloxacin* • 400 mg orally or IV daily *Levofloxacin* • 250-750 mg orally or IV daily

Quinolones

Immuno-suppression
Antimicrobials
Cardiovascular Agents
Antiosteo-porosis Agents
Antiplatelets
Diabetes Agents
Ulcer Treatment
Diuretics
Other Agents

Brand Name	Cipro® (ciprofloxacin)
Company	Schering
Class	• Synthetic, broad-spectrum quinolone with activity against *Escherichia coli, Klebsiella pneumoniae, Klebsiella oxytoca, Enterobacter aerogenes, Enterobacter cloacae, Citrobacter diversus, Citrobacter freundii, Edwardsiella tarda, Salmonella enteritidis, Salmonella typhi, Shigella sonnei, Shigella flexneri, Proteus mirabilis, Proteus vulgaris, Providencia stuartii, Providencia rettgeri, Morganella morganii, Serratia marcescens, Yersinia enterocolitica, Pseudomonas aeruginosa, Acinetobacter calcoaceticus* subsp. *Iwoffi, Acinetobacter calcoaceticus* subsp. *anitratus, Hemophilus influenzae, Hemophilus parainfluenzae, Hemophilus ducreyi, Neisseria gonorrhoeae, Neisseria meningitidis, Moraxella catarrhalis, Campylobacter jejuni, Campylobacter coli, Aeromonas hydrophila, Aeromonas caviae, Vibrio cholerae, Vibrio parahaemolyticus, Vibrio vulnificus, Brucella melitensis, Pasteurella multocida, Legionella pneumophila, Staphylococcus* species, *Streptococcus pyogenes,* and *Streptococcus pneumoniae*
Mechanism of Action	• Interferes with DNA gyrase, an enzyme which is needed for bacterial synthesis
Indications	• Lower respiratory infection • Bone and joint infection • Skin and skin structure infection • Urinary tract infection • Infectious diarrhea • Acute sinusitis • Prostatitis • Typhoid fever • Complicated intra-abdominal infection • Gonorrhea
Contraindication	• History of hypersensitivity to ciprofloxacin and to other quinolones
Warnings	• Safety and efficacy of ciprofloxacin has not been established in children, adolescents, pregnant women, and lactating women • Convulsions have been reported in patients receiving ciprofloxacin • Serious and fatal reactions have been reported after the co-administration of ciprofloxacin and theophylline • Serious hypersensitivity reactions have occurred • Pseudomembranous colitis has been reported
Special Precautions	• Crystals of ciprofloxacin have been observed in urine of human subjects • Phototoxicity has occurred after exposure to direct sunlight • Pregnancy Category C
Adverse Reactions	• Nausea • Diarrhea • Vomiting • Abdominal pain/discomfort • Headache • Restlessness • Rash

Continued ...

Quinolones

Drug Interactions	• Theophylline • Caffeine • Antacids containing magnesium, aluminum, or calcium • Cyclosporine • Warfarin • Probenecid • Drugs that can increase QT interval
Formulations	• Tablets containing 250 mg • Capsules containing 500 mg or 750 mg • 20 mL vial containing 200 mg • 40 mL vial containing 400 mg • 100 mL flexible container containing 200 mg • 200 mL flexible container containing 400 mg • 5% oral suspension • 10% oral suspension
Dosage	*Tablets* • 250 mg every 12 hours for urinary tract infection; patients with complicated infections may require 500 mg every 12 hours • A single 250 mg dose for urethral and cervical gonococcal infections • 500 mg every 12 hours for lower respiratory infection, skin and skin structure infection, infectious diarrhea, and bone and joint infection; serious infection may require 750 mg every 12 hours • Duration of therapy is usually 7 to 14 days; bone and joint infections may require therapy for 4 to 6 weeks; infectious diarrhea may be treated for 5 to 7 days *IV Infusion* • 200 mg every 12 hours for mild to moderate urinary tract infections; 400 mg every 12 hours for severe urinary tract infections • 400 mg every 12 hours for lower respiratory infection, skin and skin structure infection, and bone and joint infection • Duration of therapy is usually 7 to 14 days; bone and joint infections may require therapy for 4 to 6 weeks

Cephalosporins

Immuno-suppression

Antimicrobials

Cardiovascular Agents

Antiosteo-porosis Agents

Antiplatelets

Diabetes Agents

Ulcer Treatment

Diuretics

Other Agents

Brand Name	Ancef® (cefazolin)	Cefotan® (cefotetan)	Rocephin® (ceftriaxone)	Maxipime® (cefepime)
Company	Apothecon	Zeneca Pharmaceuticals	Roche Pharmaceuticals	Dura
Class	• Semisynthetic, broad-spectrum cephalosporin • All have activity against *Escherichia coli*, *Proteus mirabilis*, *Klebsiella* species, *Hemophilus influenzae*, *Staphylococcus aureus*, *Staphylococcus epidermidis*, and *Streptococcus pneumoniae* • Cefotan also has activity against *Morganella morganii*, *Neisseria gonorrhoeae*, *Moraxella catarrhalis*, *Proteus vulgaris*, *Providencia rettgeri*, *Serratia marcescens*, *Streptococcus agalactiae*, *Streptococcus pyogenes*, *Bacteroides bivius*, *Bacteroides disiens*, *Bacteroides fragilis*, *Bacteroides melaninogenicus*, *Bacteroides vulgatus*, *Fusobacterium* species, and gram-positive bacilli • Ceftriaxone and Cefepime also have activity against *Moraxella catarrhalis*, *Morganella morganii*, *Neisseria meningitidis*, *Serratia marcescens*, *Pseudomonas aeruginosa*, *Bacteroides* species, *Clostridium* species, *Peptostreptococcus* species, *citrobacter*, *Providencia*, *Salmonella*, and *Shigella* species			
Mechanism of Action	• Inhibition of cell wall synthesis			
Indications	• Respiratory infections • Bone and joint infection • Skin and skin structure infection • Urinary tract infection • Biliary tract infection • Genital infection • Septicemia • Endocarditis • Intra-abdominal infection • Surgical prophylaxis • Empiric therapy for febrile neutropenic patients			
Contraindication	• Hypersensitivity to cephalosporin class of antibiotics			
Warnings	• If patient is penicillin-sensitive, caution should be used because there have been reports of cross-hypersensitivity among beta-lactam antibiotics • Pseudomembranous colitis has been reported			
Special Precautions	• Patients should be monitored for signs of superinfection • Pregnancy Category B			
Adverse Reactions	• Diarrhea, nausea • Eosinophilia, positive Coombs' test, thrombocytosis, agranulocytosis, hemolytic anemia, leukopenia, thrombocytopenia, prolonged prothrombin time • Elevation of hepatic enzymes • Rash, anaphylactic reaction • Phlebitis at injection site			

Continued ...

Cephalosporins

Drug Interactions	• Aminoglycosides • Probenecid • False-positive reaction for glucose in the urine • False increases in serum and urine creatinine	
Formulations	*Cefazolin* • Vials equivalent to 500 mg or 1 g of cefazolin • Piggyback Vial equivalent to 1 g of cefazolin • Pharmacy Bulk Vial equivalent to 5 g or 10 g of cefazolin • Frozen, sterile, nonpyrogenic solution in plastic containers equivalent to 500 mg or 1 g of cefazolin *Cefotetan* • 1 g in 10 mL vial • 2 g in 20 mL vial • 1 g in 100 mL vial • 2 g in 100 mL vial • 10 g in 100 mL vial *Ceftriazone* • 10 g bulk package • 250 mg vials • 500 mg vials • 1 g vials • 2 g vials • 1 g piggyback bottles • Kit containing 500 mg ceftriaxone vial with 2.1 mL of 1% lidocaine vial for IM use • Kit containing 1 g ceftriaxone vial with 2.1 mL of 1% lidocaine vial for IM use • 1 g Add-Vantage vial • 2 g Add-Vantage vial • 1 g/50 mL premixed container • 2 g/50 mL premixed container *Cefepime* • 500 mg/15 mL vial • 1 g/100 mL piggyback vial • 1 g Add-Vantage vial • 1 g 15 mL vial • 2 g/100 mL piggyback vial • 2 g Add-Vantage vial • 2 g 20 mL vial	
Dosage	*Cefazolin* • Treatment	

Cefazolin
• Treatment

	• Moderate/severe	500 mg to 1 g every 6 to 8 hours
	• Mild	250 mg to 500 mg every 8 hours
	• UTIs	1 g every 12 hours
	• Pneumonia	500 mg every 12 hours
	• Severe/life-threatening	1 g to 1.5 g every 6 hours
• Prophylaxis		1 g IV or IM administered once 30 to 60 minutes prior to surgery; 500 mg to 1 g IV or IM may be given intraoperatively during lengthy procedures; 500 mg to 1 g IV or IM every 6 to 8 hours for 24 hours postop

Continued ...

Sidebar tabs: Immuno-suppression | Antimicrobials | Cardiovascular Agents | Antiosteo-porosis Agents | Antiplatelets | Diabetes Agents | Ulcer Treatment | Diuretics | Other Agents

Cephalosporins

Immuno-suppression

Antimicrobials

Cardiovascular Agents

Antiosteo-porosis Agents

Antiplatelets

Diabetes Agents

Ulcer Treatment

Diuretics

Other Agents

Cefotetan
- Treatment — 1 g or 2 g IV or IM every 12 hours for 5 to 10 days
- Prophylaxis — 1 g or 2 g IV administered once 30 to 60 minutes prior to surgery

Ceftriaxone
- Treatment — 1-2 g IV or IM once or twice daily for 4 to 14 days
- Prophylaxis — 1 g IV 30 to 120 minutes prior to surgery

Cefepime
- Treatment — 1-2 g IV q 12 hours for 10 days

Creatinine Clearance (mL/min)	Recommended maintenance schedule			
>60 Normal recommended dosing schedule	500 mg q12hr	1 g q12hr	2 g q12hr	2 g q8hr
30-60	500 mg q24hr	1 g q24hr	2 g q24hr	2 g q12hr
11-29	500 mg q24hr	500 mg q24hr	1 g q24hr	2 g q24hr
<11	250 mg q24hr	250 mg q24hr	500 mg q24hr	1 g q24hr

Aminoglycosides

Brand Name	Amikin® (amikacin)	Nebcin® (tobramycin)	Garamycin® (gentamicin)
Company	Apothecon	Eli Lilly and Company	Schering Corporation
Class	• Semisynthetic aminoglycoside antibiotics with activity against *Pseudomonas* species, *Escherichia coli*, *Proteus* species, *Providencia* species, *Klebsiella* species, *Enterobacter* species, *Serratia* species, *Acinetobacter* species, *Citrobacter freundii*, *Staphylococcus* species • Aminoglycosides generally have a low level of activity against gram-positive organisms		
Mechanism of Action	• Inhibit normal protein synthesis in susceptible organisms		

Indications	• Treatment of the following serious infections due to susceptible gram-negative bacteria:
	  • Septicemia
	  • Respiratory tract
	  • Bone and joint
	  • Central nervous system
	  • Skin and soft tissue
	  • Intra-abdominal
	  • Burns and postoperative infection
	  • Complicated, recurrent UTIs (aminoglycosides are not indicated for uncomplicated, initial episodes of UTIs)		

Contraindication	• History of hypersensitivity or serious toxic reaction to aminoglycosides		
Warnings	• Patients should be monitored closely because of the potential for development of ototoxicity and nephrotoxicity • Should be used with caution in premature and neonatal infants because of their renal immaturity • Neuromuscular blockade and respiratory paralysis have been reported, and the possibility of their occurrence should be considered in patients receiving anesthetics, neuromuscular blocking agents, or in patients receiving citrate-anticoagulated blood • May cause fetal damage when administered to pregnant women • Bisulfite component may cause allergic reactions • Concurrent use of potent diuretics should be avoided		
Special Precautions	• Serum and urine specimens should be collected for periodic monitoring during therapy • Pregnancy Category D • Concurrent and serial use of other ototoxic, nephrotoxic, and neurotoxic agents (systemic or topical) should be avoided • Increased nephrotoxicity has been reported after the concomitant use of cephalosporins • Patients should be well hydrated to minimize chemical irritation of the renal tubules • Dosage should be reduced if there are signs of renal dysfunction; therapy should be discontinued if azotemia increases or if there is a progressive decrease in urine output • It is especially important to monitor renal function in elderly patients • Should be used with caution in patients with neuromuscular disorders		

Continued ..

Aminoglycosides

Adverse Reactions	• *Neurotoxicity/Ototoxicity:* Hearing loss and/or loss of balance, dizziness, tinnitus, and roaring in the ears • *Neurotoxicity/Neuromuscular blockage:* Acute muscular paralysis and apnea • *Nephrotoxicity:* Elevation of serum creatinine, albuminuria, presence of red and white cells, casts, azotemia, and oliguria • *Other:* Skin rash, drug fever, headache, paresthesia, tremor, nausea/vomiting, eosinophilia, arthralgia, anemia, and hypotension
Drug Interactions	• Bacitracin • Cisplatin • Amphotericin B • Cephaloridine • Paromomycin • Viomycin • Polymyxin B • Colistin • Vancomycin • Other aminoglycosides • Potent diuretics (ethacrynic acid) • Beta-lactam antibiotics
Formulations	*Amikacin* • 100 mg/2 mL vial • 500 mg/2 mL vial • 500 mg/2 mL disposable syringe • 1 g/4 mL vial *Gentamicin* • 40 mg/mL in 2 mL vial • 40 mg/mL in 1.5 mL vial *Tobramycin* • 80 mg/2 mL vial • 40 mg/mL vial • 20 mg/2 mL vial • 60 mg/1.5 mL disposable syringe • 80 mg/2 mL disposable syringe • 60 mg/6 mL Add-Vantage vial • 80 mg/8 mL Add-Vantage vial
Dosage	*Amikacin—Patients with Normal Renal Function* • 15 mg/kg/d IM or IV divided into 2 or 3 equal doses administered at equally divided intervals • Usual duration of treatment is 7 to 10 days and it is desirable to limit the duration to the shortest time possible *Tobramycin/Gentamicin—Patients with Normal Renal Function* • 3 mg/kg/d IM or IV divided into 3 equal doses every 8 hours • Patients with life-threatening infections may be given up to 5 mg/kg/d in 3 or 4 equal doses, with a reduction to 3 mg/kg/d as soon as clinically indicated • Usual duration of treatment is 7 to 10 days and it is desirable to limit the duration to the shortest time possible

Editors' Notes:

In a septic renal transplant recipient, a loading dose of aminoglycoside may be warranted despite concerns about renal toxicity. Once daily dosing may decrease toxicity.

Immuno-suppression / Antimicrobials / Cardiovascular Agents / Antiosteoporosis Agents / Antiplatelets / Diabetes Agents / Ulcer Treatment / Diuretics / Other Agents

Linezolid

Immuno-suppression

Antimicrobials

Cardiovascular Agents

Antiosteo-porosis Agents

Antiplatelets

Diabetes Agents

Ulcer Treatment

Diuretics

Other Agents

Brand Name	**Zyvox®**
Company	Pharmacia & Upjohn Company
Class	• Oxazolidinone, synthetic antibacterial agent
Mechanism of Action	• Binds to a site on the bacterial 23S ribosomal RNA of the 50S subunit and prevents the formation of a functional 70S initiation complex, which is an essential component of the bacterial translation process
Indication	• Vancomycin-resistant *Enterococcus faecium* infections, including cases with cocurrent bacteremia • Nosocomial pneumonia caused by *Staphylococcus aureus* (methicillin-susceptible and -resistant strains), or *Streptococcus pneumoniae* (penicillin-susceptible strains only) • Complicated skin and skin structure infections caused by *Staphylococcus aureus* (methicillin-susceptible and -resistant strains), *Streptococcus pyogenes*, or *Streptococcus agalactiae* • Uncomplicated skin and skin structure infections caused by *Staphylococcus aureus* (methicillin-susceptible strains only) or *Streptococcus pyogenes* • Community-acquired pneumonia caused by *Streptococcus pneumoniae* (penicillin-susceptible strains only), including cases with concurrent bacteremia, or *Staphylococcus aureus* (methicillin-susceptible strains only)
Contraindication	• Hypersensitivity to linezolid or any other of the product components
Warning	• Myelosuppression (including anemia, leukopenia, pancytopenia, and thrombocytopenia) • Pseudomembranous colitis
Special Precautions	• Risk of superinfection • Pregnancy Category C
Adverse Reactions	• Diarrhea • Dizziness • Headache • Nausea • Vomiting • Insomnia • Constipation • Rash • Increased liver function tests
Drug Interactions	• Monoamine oxidase inhibitors • Adrenergic agents • Pseudoephedrine • Phenylpropanolamine
Formulations	*Infusion bags available in:* • 100 mL bag (200 mg linezolid) • 200 mL bag (400 mg linezolid) • 300 mL bag (600 mg linezolid) *Tablets* • 400 mg in bottles of 20 and 100 tablets, Unit dose packages of 30 tablets • 600 mg in bottles of 20 and 100 tablets, Unit dose packages of 30 tablets *Oral suspension* • 100 mg/5 mL in 240-mL glass bottles
Dosage	• 600 mg every 12 hours orally or intravenously

Editors' Notes:

Due to concerns about inappropriate use of antibiotics leading to an increase in resistant organisms, prescribers should carefully consider alternatives before initiating treatment with linezolid.

Tigecycline

Brand Name	Tygacil®
Company	Wyeth Pharmaceuticals
Class	• A broad spectrum glycylcycline antibacterial agent • Active against *Staphylococcus aureus* (methicillin-susceptible and -resistant isolates), *Staphylococcus epidermidis* (methicillin-susceptible and -resistant isolates), *Streptococcus agalactiae*, *Streptococcus pyogenes*, *Enterococcus faecalis* (vancomycin-susceptible and -resistant stains), *Enterococcus faecium* (vancomycin- susceptible and -resistant stains), *Staphylococcus haemolyticus*, *Streptococcus anginosus* grp. (includes *S. anginosus*, *S. intermedius*, and *S. constellatus*), *Citrobacter freundii*, *Enterobacter cloacae*, *Escherichia coli*, *Klebsiella oxytoca*, *Klebsiella pneumoniae*, *Bacteroides fragilis*, *Bacteroides thetaiotaomicron*, *Bacteroides uniformis*, *Bacteroides vulgatus*, *Clostridium perfringens*, *Peptostreptococcus micros*, *Enterococcus avium*, *Enterococcus casseliflavus*, *Enterococcus gallinarum*, *Listeria monocytogenes*, *Acinetobacter baumannii*, *Aeromonas hydrophila*, *Citrobacter koseri*, *Enterobacter aerogenes*, *Pasteurella multocida*, *Serratia marcescens*, *Stenotrophomonas maltophilia*, *Bacteroides distasonis*, *Bacteroides ovatus*, *Peptostreptococcus* spp., *Porphyromonas* spp., *Prevotella* spp., *Mycobacterium abscessus*, *Mycobacterium chelonae*, and *Mycobacterium fortuitum*
Mechanism of Action	• Bacteriostatic by inhibiting protein translation in bacteria by binding to the 30S ribosomal subunit and blocking entry of amino-acyl tRNA molecules into the A site of the ribosome and preventing incorporation of amino acid residues into elongating peptide chains
Indications	• Treatment of infections caused by susceptible strains of the designated microorganisms in the following conditions for patients 18 years of age and older • Complicated skin and skin structure infections caused by *Escherichia coli*, *Enterococcus faecalis* (vancomycin-susceptible isolates only), *Staphylococcus aureus* (methicillin-susceptible and -resistant isolates), *Streptococcus agalactiae*, *Streptococcus anginosus* grp. (includes *S. anginosus*, *S. intermedius*, and *S. constellatus*), *Streptococcus pyogenes* and *Bacteroides fragilis* • Complicated intra-abdominal infections caused by *Citrobacter freundii*, *Enterobacter cloacae*, *Escherichia coli*, *Klebsiella oxytoca*, *Klebsiella pneumoniae*, *Enterococcus faecalis* (vancomycin-susceptible isolates only), *Staphylococcus aureus* (methicillin-susceptible isolates only), *Streptococcus anginosus* grp. (includes *S. anginosus*, *S. intermedius*, and *S. constellatus*), *Bacteroides fragilis*, *Bacteroides thetaiotaomicron*, *Bacteroides uniformis*, *Bacteroides vulgatus*, *Clostridium perfringens*, and *Peptostreptococcus micros*
Contraindications	• Known hypersensitivity to tigecycline
Warnings	• May cause fetal harm if administered to a pregnant woman • May cause permanent teeth discoloration if used during tooth development period • Pseudomembranous colitis has been reported

Continued ...

Immuno-suppression

Antimicrobials

Cardiovascular Agents

Antiosteo-porosis Agents

Antiplatelets

Diabetes Agents

Ulcer Treatment

Diuretics

Other Agents

Tigecycline

Immunosuppression

Antimicrobials

Cardiovascular Agents

Antiosteoporosis Agents

Antiplatelets

Diabetes Agents

Ulcer Treatment

Diuretics

Other Agents

Special Precautions	• Risk of superinfection • Should be avoided during tooth development (women in the last half of pregnancy, infants, and children up to age 8) due to potential of teeth discoloration • Pregnancy Category D
Adverse Reactions	• Photosensitivity • Anti-anabolic effects • Discoloration of teeth • Diarrhea/nausea/vomiting/abdominal pain • Leukocytosis • Thrombocythemia • Elevated liver function tests • Headache • Pseudotumor cerebri • Elevation in blood urea nitrogen level • Fever
Drug Interactions	• Oral contraceptives • Warfarin
Formulation	• 5 mL glass vials containing 50°mg lyophilized powder for reconstitution
Dosage	• Initial dose 100 mg intravenously, followed by 50 mg intravenously every 12 hours for 5-14 days *Severe hepatic impairment* • Initial dose 100 mg intravenously, followed by 25 mg intravenously every 12 hours for 5-14 days

Daptomycin

Brand Name	Cubicin®
Company	Cubist Pharmaceuticals, Inc.
Class	• Cyclic lipopeptide antibacterial active against gram positive organisms • Active against *Staphylococcus aureus, Staphylococcus epidermidis, Streptococcus agalactiae, Streptococcus pyogenes, Streptococcus dysgalactiae* subsp. *Equisimilis, Enterococcus faecalis* (including vancomycin-resistant stains), *Enterococcus faecium* (including vancomycin-resistant stains), *Corynebacterium jeikeium, and Staphylococcus haemolyticus*
Mechanism of Action	• Causes rapid depolarization of membrane potential which leads to inhibition of protein, DNA and RNA synthesis and bacterial cell death
Indications	• Complicated skin and skin structure infections caused by susceptible strains of the following Gram-positive microorganisms: *Staphylococcus aureus* (including methicillin-resistant strains), *Streptococcus pyogenes, Streptococcus agalactiae, Streptococcus dysgalactiae* subsp. *equisimilis* and *Enterococcus faecalis* (vancomycin-susceptible strains only).
Contraindications	• Known hypersensitivity to daptomycin
Warnings	• Pseudomembranous colitis has been reported
Special Precautions	• Require dose adjustment in patients with renal impairment • Pregnancy Category B
Adverse Reactions	• Rash (4.3%) • Pruritus (2.8%) • Eczema • Diarrhea/nausea/vomiting/constipation • Hypotension (2.4%) • Hypertension (1.1%) • Anemia • Elevated liver function tests • Arthralgia/muscle cramps/myalgia • Rhabdomyolysis (case reports) • Increased creatine kinase level • Headache/dizziness • Insomnia • Nephrotoxicity • Injection site reactions
Drug Interactions	• Potential inhibition of metabolism of HMG CoA reductase inhibitors and increase risk of myopathy
Formulation	• 500 mg single use vials
Dosage	*Creatinine clearance ≥ 30 mL/min* • 4 mg/kg IV every 24 hours *Creatinine clearance<30 mL/min* • 4 mg/kg IV every 48 hours

Editors' Notes:

A major advance in the treatment of vancomycin-resistant enterococcus (VRE).

Antimicrobials

Cardiovascular Agents

Antiosteo-porosis Agents

Antiplatelets

Diabetes Agents

Ulcer Treatment

Diuretics

Other Agents

Antifungals Overview

Agent	Dosage
Nystatin Suspension	• 400,000 to 600,000 units 4 times a day, continued until 48 hours after symptoms have disappeared
Clotrimazole Troche	*Treatment* • One troche 5 times a day for 14 consecutive days *Prophylaxis* • One troche 3 times a day until steroids are reduced to maintenance levels
Fluconazole	*Oropharyngeal Candidiasis* • 200 mg on first day, followed by 100 mg/d for at least 2 weeks *Esophageal Candidiasis* • 200 mg on first day, followed by 100 mg/d for a minimum of 3 weeks and for at least 2 weeks after resolution of symptoms *Systemic Candidiasis* • Up to 400 mg/d *Cryptococcal Meningitis* • 400 mg on first day, followed by 200 mg/d for 10 to 12 weeks after cerebrospinal fluid becomes culture negative
Posaconazole	• 200 mg orally three times daily with a full meal or with a liquid nutritional supplement
Voriconazole	*Treatment of Esophageal Candidiasis* • 200 mg orally every 12 hours *Invasive Aspergillosis, Candidemia in non-neutropenic patients and other deep tissue Candida infection, and Scedosporiosis and Fusariosis* • Loading dose 6 mg/kg intravenously every 12 hours for the first 24 hours • Maintenance dose 4 mg/kg intravenously every 12 hours or 200 mg orally twice daily
Amphotericin B	*IV Infusion* • Initially 1 mg (base) as a test dose, administered in 20 mL of 5% dextrose injection over a period of 20 to 30 minutes, with 5 mg to 10 mg incremental increases depending on patient tolerance and severity of the infection, up to a maximum of 0.5 to 0.7 mg/kg • Should be infused over a period of 2 to 6 hours
Amphotericin B Lipid Complex Injection	• 5 mg/kg/d IV for invasive aspergillosis • 2.0 mg/kg/d to 2.5 mg/kg/d IV for candidiasis
Miconazole Vaginal Cream 2%	*Tinea Pedis, Tinea Cruris, Tinea Corporis, and Cutaneous Candidiasis* • Enough to cover affected areas twice daily (morning and evening) *Tinea Versicolor* • Enough to cover affected areas once daily
Ketoconazole	*Adults* • 200 mg in a single daily administration • If there is insufficient response, dose can be raised to 400 mg/d *Children* • 3.3 mg/kg to 6.6 mg/kg in a single daily dose

Continued ...

Immuno-suppression

Antimicrobials

Cardiovascular Agents

Antiosteo-porosis Agents

Antiplatelets

Diabetes Agents

Ulcer Treatment

Diuretics

Other Agents

Antifungals Overview

Agent	Dosage
Caspofungin	• 70 mg loading dose followed by 50 mg daily intravenously
Micafungin	*Treatment of Esophageal Candidiasis* • 150 mg intravenously every 24 hours *Prophylaxis of Candida Infections* • 50 mg intravenously every 24 hours
Anidulafungin	*Esophageal Candiditis* • 100 mg intravenously loading dose on Day 1, followed by 50 mg intravenously daily *Candidemia and Other Disseminated Candidiasis* • 200 mg intravenously loading dose on Day 1, followed by 100 mg intravenously every 24 hours

Editors' Notes:

Several new antifungal agents have been approved. Caspofungin and voriconazole have efficacy against aspergillus infection and are not nephrotoxic. The utility of these agents has led to improved survival in transplant recipients infected with this devastating fungus.

Immuno-suppression

Antimicrobials

Cardiovascular Agents

Antiosteo-porosis Agents

Antiplatelets

Diabetes Agents

Ulcer Treatment

Diuretics

Other Agents

Nystatin Suspension

Brand Name	• **Mycostatin®**
Company	• Bristol-Myers Oncology • Generics
Class	• Antifungal with activity against *Candida albicans* and a wide variety of yeasts
Mechanism of Action	• Binds to sterols in fungus cell membrane, resulting in an inability to function as a selective barrier and a subsequent loss of essential cellular components
Indication	• Treatment and prophylaxis of oropharyngeal candidiasis
Contraindication	• Hypersensitivity to nystatin
Special Precautions	• Patients with full or partial dentures may need to soak their dentures nightly in Mycostatin (nystatin) • Pregnancy Category C
Adverse Reactions	• Diarrhea • Nausea/vomiting • Stomach pain • Rash
Drug Interactions	• None reported
Formulations	*Nystatin Oral Suspension* • 100,000 units/mL A teaspoon equals approximately 500,000 units
Dosage	• 400,000 or 600,000 units 4 times a day, continued until 48 hours after symptoms have disappeared

Editors' Notes:

Effective prophylaxis for oral candidiasis, especially in diabetic patients.

Antimicrobials

Cardiovascular Agents

Antiosteo-porosis Agents

Antiplatelets

Diabetes Agents

Ulcer Treatment

Diuretics

Other Agents

Clotrimazole Troche

Brand Name	**Mycelex®**
Company	Bayer Corporation
Class	• Synthetic, broad-spectrum antifungal agent with activity against *Candida albicans* and other *Candida* species
Mechanism of Action	• Inhibits growth of pathogens by altering permeability of cell membranes
Indications	• Local treatment of oropharyngeal candidiasis • Prophylaxis to reduce incidence of oropharyngeal candidiasis in immunocompromised patients
Contraindication	• Hypersensitivity to clotrimazole
Warning	• Is not indicated for the treatment of systemic mycoses
Special Precautions	• Abnormal liver function tests have been reported • Patients must be of age where they can understand the need to let each troche dissolve slowly in the mouth • Pregnancy Category C
Adverse Reactions	• Abnormal liver function test values • Nausea/vomiting • Unpleasant mouth sensations • Pruritus
Drug Interactions	• None reported
Formulation	• Slowly dissolving tablet containing 10 mg of clotrimazole, supplied in bottles of 70 and 140 tablets
Dosage	*Treatment* • One troche 5 times a day for 14 consecutive days *Prophylaxis* • One troche 3 times a day until steroids are reduced to maintenance levels

Editors' Notes:

Better compliance than with nystatin. Currently, the drug of choice for prophylaxis. A report suggested that clotrimazole troche can increase tacrolimus blood levels. (Vasquez E, Pollak R, Benedetti E. Clotrimazole increases tacrolimus blood levels: a drug interaction in kidney transplant patients. Clin Transplant ©2001 15(2):95-9.)

Immuno-suppression | Antimicrobials | Cardiovascular Agents | Antiosteo-porosis Agents | Antiplatelets | Diabetes Agents | Ulcer Treatment | Diuretics | Other Agents

Fluconazole

Brand Name	Diflucan®
Company	Pfizer
Class	• Synthetic, broad-spectrum bis-triazole antifungal agent with activity against *Cryptococcus neoformans* and *Candida* species
Mechanism of Action	• Selectively inhibits fungal cytochrome P-450 sterol C-14 alpha-demethylation, leading to accumulation of 14 alpha-methyl sterols in fungi
Indications	• Treatment of oropharyngeal and esophageal candidiasis and may be effective for the treatment of serious systemic candidal infections • Cryptococcal meningitis
Contraindications	• Hypersensitivity to fluconazole or to any of its excipients • Caution should be used in patients with hypersensitivity to other azoles • Should not be used with terfenadine or astemizole
Warnings	• Anaphylaxis has been reported • Patients who develop abnormal liver function tests should be monitored for development of more serious hepatic injury • Immunocompromised patients who develop rashes should be monitored closely • Cardiac events including torsade de pointes have been reported in patients on concomitant cisapride therapy
Special Precaution	• Pregnancy Category C
Adverse Reactions	• Nausea • Headache • Skin rash • Vomiting • Abdominal pain • Diarrhea
Drug Interactions	• Increased prothrombin time after warfarin administration • Phenytoin • Cyclosporine • Tolbutamide • Tacrolimus • Glyburide • Glipizide • Rifampin • Cisapride • Terfenadine • Astemizole • Theophylline
Formulations	• Tablets containing 50 mg, 100 mg, or 200 mg • Sterile isotonic solutions containing 2 mg/mL of fluconazole in glass bottles or plastic containers holding volumes of 100 mL or 200 mL • Powder for oral suspension—10 mg/mL when reconstituted • Powder for oral suspension—40 mg/mL when reconstituted
Dosage	*Oropharyngeal Candidiasis* • 200 mg on first day, followed by 100 mg/d for at least 2 weeks *Esophageal Candidiasis* • 200 mg on first day, followed by 100 mg/d for a minimum of 3 weeks and for at least 2 weeks after resolution of symptoms *Systemic Candidiasis* • Up to 400 mg/d *Cryptococcal Meningitis* • 400 mg on first day, followed by 200 mg/d for 10 to 12 weeks after cerebrospinal fluid becomes culture negative *Dosage Adjustments for Renal Impairment* CrCl (mL/min) / % of Recommended Dose >50 — 100% 11-50 — 50% Patients on hemodialysis — One recommended dose after each dialysis

Editors' Notes:
Commonly affects cyclosporine level. May be useful for candida UTI and esophageal candidiasis. Co-administration of terfenadine or cisapride is contraindicated due to an increased risk of prolonged QT intervals and torsade de pointes. However, both drugs have now been removed from the market.

Immunosuppression · Antimicrobials · Cardiovascular Agents · Antiosteoporosis Agents · Antiplatelets · Diabetes Agents · Ulcer Treatment · Diuretics · Other Agents

Posaconazole

Brand Name	Noxafil®
Company	Schering
Class	• Broad-spectrum triazole antifungal agent with activity against *Aspergillus fumigatus* and *Candida albicans*
Mechanism of Action	• Selectively inhibits fungal cytochrome P-450-mediated 14 alpha-lanosterol demethylation, an essential step in fungal ergosterol biosynthesis, leading to accumulation of 14 alpha-methyl sterols and the subsequent loss of ergosterol in the fungal cell membrane
Indications	• Prophylaxis of invasive *Aspergillus* and *Candida* infections in severely immunocompromised patients aged 13 years and older
Contraindications	• Hypersensitivity to posaconazole or to any components of the product • Co-administration with terfenadine, astemizole, cisapride, or quinidine may lead to QTc prolongation and torsades de pointes • Caution should be used in patients with hypersensitivity to other azoles • Co-administration with rifampin, efavirenz, rifabutin, high-dose ritonavir, carbamazepine, long- acting barbiturates, sirolimus, astemizole, quinidine and ergot alkaloids
Warnings	• Patients who develop abnormal liver function tests should be monitored for development of more serious hepatic injury
Special Precautions	• Has been associated with arrhythmias and QT prolongation • Pregnancy Category C
Adverse Reactions	• Edema • Prolonged QT interval • Diarrhea/nausea/vomiting/abdominal pain • Dyspepsia • Xerostomia • Rash/pruritus • Electrolyte abnormalities • Hypotension/hypertension • Arthralgia/back pain • Elevated liver function tests • Confusion • Headache • Dizziness • Insomnia • Anxiety • Fatigue • Vaginitis
Drug Interactions	• Phenytoin • Benzodiazepines • Cyclosporin • HMG-CoA reductase inhibitors (statins) • Tacrolimus • Calcium channel blockers • Rifabutin • Ergot alkaloids • Sirolimus • Vinca alkaloids • Quinidine
Formulations	• Oral suspension 40 mg/mL in 4 oz amber glass bottle
Dosage	• The duration of therapy is based on recovery from neutropenia or immunosuppression • 200 mg orally three times daily with a full meal or with a liquid nutritional supplement

Voriconazole

Immuno-suppression

Brand Name	Vfend®
Company	Pfizer
Class	• Broad-spectrum triazole antifungal agent with activity against *Aspergillus* species (*A. fumigatus, A. flavus, A. niger* and *A. terreus*), *Candida* species (*C. albicans, C. glabrata, C. krusei, C. parapsilosis* and *C. tropicalis*), *Scedosporium apiospermum* and *Fusarium* spp., including *Fusarium solani*
Mechanism of Action	• Selectively inhibits fungal cytochrome P-450-mediated 14 alpha-lanosterol demethylation, an essential step in fungal ergosterol biosynthesis, leading to accumulation of 14 alpha-methyl sterols and the subsequent loss of ergosterol in the fungal cell membrane
Indications	• Invasive aspergillosis • Candidemia in nonneutropenic patients and the following *Candida* infections: disseminated infections in skin and infections in abdomen, kidney, bladder wall, and wounds • Esophageal candidiasis • Serious fungal infections caused by *Scedosporium apiospermum* (asexual form of *Pseudallescheria boydii*) and *Fusarium* spp. including *Fusarium solani,* in patients intolerant of, or refractory to, other therapy
Contraindications	• Hypersensitivity to voriconazole or to any of its excipients • Caution should be used in patients with hypersensitivity to other azoles • Coadministration with rifampin, efavirenz, rifabutin, high-dose ritonavir, carbamazepine, long- acting barbiturates, sirolimus, astemizole, quinidine and ergot alkaloids
Warnings	• Anaphylactoid reactions have been reported • Patients who develop abnormal liver function tests should be monitored for development of more serious hepatic injury • Visual disturbances have been reported
Special Precautions	• Intravenous formulation should be avoid in patients with moderate to severe renal insufficiency (creatinine clearance less than 50 milliliters/minute) due to the accumulation of the intravenous vehicle sulfobutyl ether beta-cyclodextrin sodium • Pregnancy Category D
Adverse Reactions	• Peripheral edema • Diarrhea/nausea/vomiting/abdominal pain • Rash • Hypotension/hypertension • Tachycardia • Visual disturbances—altered visual perception, blurred vision, color vision changes, and photophobia • Elevated liver function tests • Headache

Continued ...

Antimicrobials

Cardiovascular Agents

Antiosteo-porosis Agents

Antiplatelets

Diabetes Agents

Ulcer Treatment

Diuretics

Other Agents

Voriconazole

Drug Interactions	• Increased prothrombin time after warfarin administration • Phenytoin • Carbamazepine • Cyclosporin • Tacrolimus • Methadone • Oral contraceptives • Sirolimus • Efavirenz • Omeprazole • Benzodiazepines • HMG-CoA reductase inhibitors (statins) • Dihydropyridine calcium channel blockers • Sulfonylurea oral hypoglycemics • Vinca alkaloids
Formulations	• Single-use vial of 50 mg • 50 mg tablet • 200 mg tablet • 200 mg in a single use vial as a sterile lyophilized powder • Oral powder for suspension 40 mg/mL in 100 mL high density polyethylene bottles
Dosage	• Patients should be treated for at least 14 days following resolution of symptoms or following last positive culture, whichever is longer. *Treatment of Esophageal Candidiasis* • 200 mg orally every 12 hours *Invasive Aspergillosis, Candidemia in nonneutropenic patients and other deep tissue Candida infection, and Scedosporiosis and Fusariosis* • Loading dose 6 mg/kg intravenously every 12 hours for the first 24 hours • Maintenance dose 4 mg/kg intravenously every 12 hours or 200 mg orally twice daily

Amphotericin B

Immuno-suppression

Brand Name	Fungizone®
Company	Apothecon
Class	• Antifungal polyene antibiotic obtained from a strain of *Streptomyces nodosus*
Mechanism of Action	• Binds to sterols in fungus cell membrane, resulting in an inability to function as a selective barrier and a subsequent loss of essential cellular components
Indications	• Aspergillosis • Leishmaniasis • Blastomycosis • Zygomycosis • Candidiasis • Mucormycosis • Coccidioidomycosis • Conidiobolus • Cryptococcosis • Sporotrichosis • Basidiobolus • Histoplasmosis • Paracoccidioidomycosis *Life threatening fungal infections including:* • Endocarditis • Fungal urinary tract infection • Endophthalmitis • Meningoencephalitis • Intra-abdominal infection • Meningitis (cryptococcal, fungal) • Septicemia
Contraindications	• Hypersensitivity to amphotericin B • Renal function impairment
Warning	• Because of its toxicity, amphotericin B is indicated primarily in patients with progressive, potentially fatal infections in whom the diagnosis is firmly established, preferably by positive culture or histologic study

Special Precautions	• Pregnancy Category B
	• It is important to monitor the following patient parameters
	  • BUN
	  • Serum creatinine
	  • CBC
	  • Platelet count
	  • Serum magnesium and potassium

Adverse Reactions	• Anemia • Hypokalemia • Renal function impairment • Thrombophlebitis • Infusion-related illness (fever, chills, nausea, vomiting, headache, hypotension) • Blurred vision • Cardiac arrhythmias • Hypersensitivity • Leukopenia • Polyneuropathy • Seizures • Thrombocytopenia

Continued ...

Antimicrobials   Cardiovascular Agents   Antiosteo-porosis Agents   Antiplatelets   Diabetes Agents   Ulcer Treatment   Diuretics   Other Agents

Amphotericin B

Immuno-suppression

Antimicrobials

Cardiovascular Agents

Antiosteo-porosis Agents

Antiplatelets

Diabetes Agents

Ulcer Treatment

Diuretics

Other Agents

Drug Interactions	• Adrenocorticoids, glucocorticoid, mineralocorticoid, carbonic anhydrase inhibitors, corticotropin • Medications that cause blood dyscrasia or bone marrow suppression • Radiation therapy • Digitalis glycosides • Neuromuscular blocking agents • Potassium-depleting diuretics • Nephrotoxic medications • Flucytosine
Formulation	• 50 mg (base)
Dosage	*IV Infusion* • Initially 1 mg (base) as a test dose, administered in 20 mL of 5% dextrose injection over a period of 20 to 30 minutes, with 5 mg to 10 mg incremental increases depending on patient tolerance and severity of the infection, up to a maximum of 0.5 mg/kg to 0.7 mg/kg • Maximum dose 1.5 mg/kg/day • Should be infused over a period of 2 to 6 hours

Editors' Notes:

Universal concomitant nephrotoxicity with cyclosporine-A. Requires close dose monitoring. Hydration with sodium chloride prior to dosing may help prevent renal toxicity. Renal toxicity is usually reversible.

Amphotericin B Lipid Complex Injection

Brand Name	Abelcet™	Amphotec	Ambisome
Company	Enzon	Sequus Pharmaceuticals	Astellas
Class	• Antifungal polyene antibiotic obtained from a strain of *Streptomyces nodosus*		
Mechanism of Action	• Binds to sterols in fungus cell membrane, resulting in an inability to function as a selective barrier and a subsequent loss of essential cellular components		
Indication	• Fungal infections in patients who are refractory to or intolerant of conventional amphotericin B therapy.		
Contraindication	• Hypersensitivity to amphotericin B		
Warning	• Anaphylaxis has been reported with amphotericin B-containing products—facilities equipped for cardiopulmonary resuscitation should be available during administration		

Special Precautions	• Pregnancy Category B
	• It is important to monitor the following patient parameters
	  • Serum creatinine • Tacrolimus
	  • Liver function • Azoles
	  • Complete blood count
	  • Serum magnesium and potassium		

Adverse Reactions	• Infusion-related illness (fever, chills, nausea, vomiting)		
Drug Interactions	• Antineoplastic agents • Corticosteroids and corticotropin • Cyclosporine-A • Digitalis glycosides • Flucytosine • Imidazoles • Leukocyte transfusions • Nephrotoxic medications • Skeletal muscle relaxants • Zidovudine		
Formulation	• 50 mg and 100 mg vial for injection (Ambiosome only 50 mg vials)		

Dosage	• Abelcet: 5 mg/kg/d IV • Amphotec: 3-4 mg/kg/day
	• Ambisome:
	    Empiric therapy: 3 mg/kg/day
	    Systemic fungal: 3-5 mg/kg/day
	    Cryptococcal/menningitis: 6 mg/kg/day		

Editors' Notes:

Lipid and lysosomal amphotericn B may reduce the incidence of renal toxicity but are expensive.

2.0 mg/kg/d to 2.5 mg/kg/d IV for candidiasis have been used.

Ambiosome may be less toxic, but is more expensive.

Immuno-suppression

Antimicrobials

Cardiovascular Agents

Antiosteo-porosis Agents

Antiplatelets

Diabetes Agents

Ulcer Treatment

Diuretics

Other Agents

Miconazole Vaginal Cream 2%

Brand Name	**Monistat-derm®**
Company	Ortho-McNeil Pharmaceutical
Class	• Synthetic antifungal agent with activity against common dermatophytes including *Trichophyton rubrum*, *Trichophyton mentagrophytes*, and *Epidermophyton floccosum*, yeast-like fungi such as *Candida albicans*, and *Malassezia furfur*, the organism responsible for tinea versicolor
Mechanism of Action	• Inhibits growth of dermatophytes and yeast-like fungi
Indications	• Treatment of tinea pedis, tinea cruris, and tinea corporis caused by susceptible organisms • Treatment of cutaneous candidiasis • Treatment of tinea versicolor
Contraindication	• None noted
Warning	• None noted
Special Precautions	• Treatment should be discontinued if there are signs of sensitivity or chemical irritation • Cream is for external use only
Adverse Reactions	• Irritation • Burning • Maceration • Allergic contact dermatitis
Drug Interactions	• None reported
Formulation	• Cream containing miconazole nitrate at 2% strength, supplied in 15 g, 1 oz, and 3 oz tubes
Dosage	*Tinea Pedis, Tinea Cruris, Tinea Corporis, and Cutaneous Candidiasis* • Enough to cover affected areas twice daily (morning and evening) *Tinea Versicolor* • Enough to cover affected areas once daily

Ketoconazole

Brand Name	**Nizoral®**
Company	Janssen Pharmaceutica
Class	• Synthetic broad-spectrum antifungal agent with clinical activity against *Blastomyces dermatitidis*, *Candida* species, *Coccidioides immitis*, *Histoplasma capsulatum*, *Paracoccidioides brasiliensis*, *Phialophora* species, *Trichophyton* species, *Epidermophyton* species, and *Microsporum* species
Mechanism of Action	• Impairs the synthesis of ergosterol, which is a vital component of fungus cell membranes
Indications	• Treatment of the following systemic fungal infections: • Candidiasis • Chronic mucocutaneous candidiasis • Oral thrush • Candiduria • Blastomycosis • Coccidioidomycosis • Histoplasmosis • Chromomycosis • Paracoccidioidomycosis
Contraindications	• Hypersensitivity to ketoconazole • Concurrent use of terfenadine or astemizole
Warnings	• Hepatotoxicity, primarily of the hepatocellular type, has been reported • Anaphylaxis may occur after the first dose • Deaths within 2 weeks of treatment initiation have been reported in patients with prostate cancer; the role of ketoconazole in these deaths has not been ascertained, but it is known that ketoconazole can suppress adrenal corticosteroid secretion
Special Precautions	• Ketoconazole has been shown to lower serum testosterone • Drug-induced achlorhydria can cause reduction in absorption of ketoconazole • Pregnancy Category C
Adverse Reactions	• Nausea/vomiting • Abdominal pain • Pruritus
Drug Interactions	• May enhance anticoagulant effects of coumarin-like drugs • Rifampin • INH • Cyclosporine • Phenytoin • Hypoglycemic agents • Terfenadine • Astemizole
Formulations	• Tablets containing 200 mg of ketoconazole, supplied in bottles of 100 tablets and in blister packs of 10 X 10 tablets • Also available in cream and shampoo formulations
Dosage	*Adults* • 200 mg in a single daily administration • If there is insufficient response, dose can be raised to 400 mg/d *Children* • 3.3 mg/kg to 6.6 mg/kg in a single daily dose

Editors' Notes:

Ketoconazole has been used in cardiac and renal transplant recipients to lower cyclosporine doses. The long-term safety of this approach has not been established.

Caspofungin

Brand Name	Cancidas®
Company	Merck
Class	• Echinocandin antifungal agent
Mechanism of Action	• Inhibits the synthesis of (beta) (1,3)-D-glucan, an integral component of fungus cell wall
Indications	• Treatment of the invasive aspergillosis in patients who are refractory to or intolerant of other therapies (i.e., amphotericin B, lipid formulations of amphotericin B, and/or itraconazole).
Contraindications	• Hypersensitivity to any component of this product.
Warnings	• Concomitant use of caspofungin with cyclosporine is not recommended unless the potential benefit outweighs the potential risk to the patient.
Special Precautions	• Pregnancy Category C
Adverse Reactions	• Nausea • Facial swelling • Pruritus • Vomiting • Anaphalaxis • Rash • Fever • Increased ALT • Increased AST • Decreased hematocrit • Decreased hemoglobin • Increased serum alkaline phosphatase • Infused vein complications
Drug Interactions	• Inducers of drug clearance and/or mixed inducers/inhibitors • Cyclosporine • Tacrolimus
Formulations	• Single-use vial of 50 mg and 70 mg
Dosage	A single 70-mg loading dose should be administered on Day 1, followed by 50 mg daily, administered by slow IV infusion of approximately 1 hour. Duration of treatment should be based on the severity of patient's underlying disease, recovery from immunosuppression, and clinical response.

Editors' Notes:

Caspofungin has good activity against candida, and is currently being reviewed by the FDA for this indication. In two clinical studies, cyclosporine (one 4 mg/kg dose or two 3 mg/kg doses) increased the AUC of caspofungin by approximately 35%. Cancidas did not increase the plasma levels of cyclosporine. There were transient increases in liver ALT and AST when caspofungin and cyclosporine were coadministered.

Micafungin

Brand Name	**Mycamine®**
Company	Astellas Pharmaceuticals
Class	• Echinocandin antifungal agent
Mechanism of Action	• Interferes with the formation of fungal cell wall by inhibiting the synthesis of (1,3)-beta-D-glucan
Indications	• Treatment of patients with esophageal candidiasis • Prophylaxis of *Candida* infections in patients undergoing hematopoietic stem cell transplantation
Contraindications	• Hypersensitivity to any components of this product
Warnings	• Isolated cases of anaphylactoid reactions have been reported
Special Precautions	• Pregnancy Category C
Adverse Reactions	• Flushing • Hypotension • Phlebitis • Rash • Pruritus • Diarrhea/nausea/vomiting/abdominal pain • Anemia • Leukopenia • Thrombocytopenia • Neutropenia • Hyperbilirubinemia • Elevated liver function tests • Headache/dizziness • Somnolence • Hypocalcemia • Hypokalemia • Hypomagnesemia • Elevated creatinine and blood urea nitrogen level • Fever/rigor • Histamine-mediated symptoms • Injection site pain
Drug Interactions	Increase exposure of the following drugs • Nifedipine • Sirolimus
Formulations	• Single-use vial of 50 mg
Dosage	*Treatment of Esophageal Candidiasis* • 150 mg intravenously every 24 hours *Prophylaxis of Candida Infections* • 50 mg intravenously every 24 hours

Immuno-suppression

Antimicrobials

Cardiovascular Agents

Antiosteo-porosis Agents

Antiplatelets

Diabetes Agents

Ulcer Treatment

Diuretics

Other Agents

Anidulafungin

Immuno-suppression

Antimicrobials

Cardiovascular Agents

Antiosteo-porosis Agents

Antiplatelets

Diabetes Agents

Ulcer Treatment

Diuretics

Other Agents

Brand Name	Eraxis®
Company	Pfizer
Class	• Echinocandin antifungal agent
Mechanism of Action	• Interferes with the formation of fungal cell wall by inhibiting the synthesis of (1,3)-beta-D-glucan
Indications	• Treatment of patients with esophageal candidiasis • Treatment of candidemia and other forms of *Candida* infections (intra-abdominal abscess, and peritonitis)
Contraindications	• Hypersensitivity to any components of this product or any other echinocandins
Warnings	• Hepatic dysfunction, worsening hepatic failure, hepatitis, or clinically significant hepatic abnormalities have been reported
Special Precautions	• Pregnancy Category C
Adverse Reactions	• Flushing • Hypotension • Rash • Pruritus • Diarrhea/nausea • Neutropenia • Elevated liver function tests • Headache • Hypokalemia • Histamine-mediated symptoms
Drug Interactions	Cyclosporine increases exposure of the anidulafungin
Formulations	• Single-use vials of 50 mg
Dosage	*Esophageal Candidiasis* • 100 mg intravenously loading dose on Day 1, followed by 50 mg intravenously daily *Candidemia and Other Disseminated Candidiasis* • 200 mg intravenously loading dose on Day 1, followed by 100 mg intravenously every 24 hours

Chapter 3
Cardiovascular Agents

Cardiovascular Agents Overview

Calcium Channel Blockers

- Diltiazem
- Nifedipine
- Isradipine
- Verapamil
- Amlodipine
- Amlodipine and Benazepril Hydrochloride

Alpha-Adrenergic Receptor Blockers

- Terazosin
- Doxazosin
- Labetalol

Central Alpha-Adrenergic Agonist

- Clonidine

Beta-Blockers

- Metoprolol
- Atenolol
- Carvedilol

ACE Inhibitors

- Lisinopril
- Enalapril
- Captopril
- Perindopril
- Quinapril
- Ramipril
- Benazepril
- Trandolapril
- Fosinopril
- Moexipril

Angiotensin II Receptor Antagonists

- Losartan
- Valsartan
- Candesartan
- Irbesartan
- Telmisartan
- Eprosartan

Cholesterol-Lowering Agents

- Lovastatin
- Pravastatin
- Fluvastatin
- Simvastatin
- Atorvastatin
- Rosuvastatin
- Bile Acid Sequestrants
- Ezetimibe

Cardiovascular Agents Overview

Agent	Dosage
Calcium Channel Blockers	• Dosage should be individualized by titration *Diltiazem* • 180 mg/d to 240 mg/d *Nifedipine* • 30 mg/d or 60 mg/d *Isradipine* • 2.5 mg BID alone or in combination with a thiazide diuretic *Verapamil* • 180 mg/d with upward titrations if adequate response is not obtained *Amlodipine* • 5 mg/d with a maximum dose of 10 mg/d *Amlodipine and benazepril hydrochloride* • 2.5 mg to 5 mg of amlodipine and 10 mg to 20 mg of benazepril once daily
Alpha-Adrenergic Receptor Blockers	• Dosage must be individualized *Terazosin* • Initial dosage is 1 mg at bedtime • Maintenance range is usually 1 mg/d to 5 mg/d *Doxazosin* • Initial dosage is 1 mg given once daily • Depending on standing blood pressure, dose may be increased to 2 mg and then if necessary to 4 mg, 8 mg, or 16 mg *Labetalol (α- and β-blocker)* • 100 mg twice daily • After 2 or 3 days, dosage may be titrated in increments of 100 mg BID every 2 or 3 days; usual maintenance range is 200 mg to 400 mg twice daily
Central Alpha-Adrenergic Agonist	• Dosage must be individualized *Clonidine* • Initial dosage is 0.1 mg twice a day • Maintenance dose is reached with 0.1 mg/d increments until patient reaches the appropriate maintenance range of 0.2 mg/d to 0.6 mg/d given in divided doses
Beta-Blockers	• Dosage must be individualized *Metoprolol* • Initial dosage is 100 mg/d in single or divided doses alone or added to diuretic therapy • Dosage can be increased at weekly (or longer intervals) until optimal control is achieved, usually at 100 mg/d to 450 mg/d *Atenolol* • Initial dosage is 50 mg/d given as 1 tablet alone or added to diuretic therapy • If optimal effect is not achieved within 1 to 2 weeks, dosage can be increased to 100 mg/d given as a single tablet *Carvedilol* • 3.125 mg to 50 mg orally twice daily

Continued ...

Side tab index (right margin): Immuno-suppression | Antimicrobials | Cardiovascular Agents | Antiosteo-porosis Agents | Antiplatelets | Diabetes Agents | Ulcer Treatment | Diuretics | Other Agents

Cardiovascular Agents Overview

Agent	Dosage
ACE inhibitors	• Dosage should be individualized *Lisinopril* • Initial dosage is 10 mg/d given once a day • Usual maintenance range is 20 mg/d to 40 mg/d in a single dose *Enalapril* • Initial dosage is 5 mg/d given once a day • Usual maintenance range is 10 mg/d to 40 mg/d in a single dose or in 2 divided doses *Captopril* • Initial dosage is 25 mg BID or TID • If optimal effect is not achieved within 1 to 2 weeks, dosage can be increased to 50 mg BID or TID *Perindopril* • 4 to 8 mg daily *Quinapril* • 10 to 80 mg daily or divided twice daily *Ramipril* • 2.5 to 20 mg daily or divided twice daily *Benazepril* • 5 to 40 mg daily *Trandolapril* • 1 to 8 mg daily *Fosinopril* • 5 to 40 mg daily *Moexipril* • 3.75 to 60 mg daily
Angiotensin II Receptor Antagonists	*Losartan* • 50 mg in patients daily • 25 mg in patients who may be volume depleted *Valsartan* • 80 mg once daily • Dosage range from 80 mg to 320 mg *Candesartan* • 8 to 32 mg daily *Irbesartan* • 150 to 300 mg daily *Telmisartan* • 20 to 80 mg daily *Eprosartan* • 400 to 800 mg daily *Olmesartan Medoxomil* • 20 to 40 mg daily

Continued ...

Cardiovascular Agents Overview

Cholesterol-Lowering Agents	• Dosage should be individualized

• Dosage should be individualized

Lovastatin
• Initial dosage is 20 mg/d taken once daily with the evening meal
• Patients on immunosuppressive therapy should be started on 10 mg/d and should not exceed 20 mg/d
• Maximum recommended dose is 80 mg/d

Pravastatin
• Starting dose is 10 mg/d or 20 mg/d taken once daily at bedtime
• Patients on immunosuppressive therapy should be started on 10 mg/d and should not exceed 20 mg/d

Fluvastatin
• Initial dose is 20 mg/d at bedtime
• Recommended dosage range is 20 mg/d to 40 mg/d as a single dose

Simvastatin
• Starting dosage range is 5 mg/d to 10 mg/d
• Maximum dosage is 40 mg/d

Atorvastatin
• Initial dose of 10 mg daily, individualized according to LDL concentrations, to a maximum of 80 mg once daily

Ezetimibe
• 10 mg orally once daily

Rosuvastatin
• 5-40 mg orally daily

Colesevelam
• 1875 mg orally twice daily

Immuno-suppression

Antimicrobials

Cardiovascular Agents

Antiosteo-porosis Agents

Antiplatelets

Diabetes Agents

Ulcer Treatment

Diuretics

Other Agents

Calcium Channel Blockers

Immuno-suppression

Antimicrobials

Cardiovascular Agents

Antiosteo-porosis Agents

Antiplatelets

Diabetes Agents

Ulcer Treatment

Diuretics

Other Agents

Brand Name	Cardizem® CD (diltiazem)	Procardia® XL (nifedipine)	DynaCirc® (isradipine)
Company	Biovail	Pfizer	Reliant
Mechanism of Action	• Modulate influx of ionic calcium across the cell membrane of arterial smooth muscle as well as in conductile and contractile myocardial cells		
Indication	• Management of essential hypertension		
Contraindications	• Hypersensitivity to diltiazem • Hypotension • Sick sinus syndrome • Second- or third-degree AV block • Acute myocardial infarction and pulmonary congestion	• Hypersensitivity to Nifedipine	• Hypersensitivity to isradipine
Adverse Reactions	• Headache • Bradycardia • Edema • Dizziness • ECG abnormalities • AV block first degree • Asthenia	• Edema • Headache • Fatigue • Dizziness • Constipation • Nausea	• Headache • Dizziness • Edema • Palpitations • Fatigue • Flushing
Drug Interactions	• Other antihypertensive agents (vasodilators, ACE inhibitors, diuretics, and beta-blockers) • Digoxin • Disopyramide • Flecainide • Quinidine • Cimetidine • Lithium	• Carbamazepine • Rifampin • Phenobarbital • Cyclosporine • Theophylline • Inhalation anesthetics • Neuromuscular blocking agents	
Dosage	• Dose should be individualized by titration • 180 mg/d to 240 mg/d	• Dose should be individualized by titration • 30 mg/d or 60 mg/d	• Dose should be individualized • 2.5 mg BID alone or in combination with a thiazide diuretic

Calcium Channel Blockers

Brand Name	Calan® SR (verapamil)	Norvasc® (amlodipine)	Lotrel™ (amlodipine and benazepril hydrochloride)
Company	G.D. Searle & Co.	Pfizer Inc.	Novartis
Mechanism of Action	• Modulate influx of ionic calcium across the cell membrane of arterial smooth muscle as well as in conductile and contractile myocardial cells		
Indication	• Management of essential hypertension		
Contraindications	• Hypersensitivity to verapamil • Severe left ventricular dysfunction • Hypotension • Sick sinus syndrome • Second- or third-degree AV block • Atrial flutter or atrial fibrillation and an accessory bypass tract	• Hypersensitivity to amlodipine	• Hypersensitivity to amlodipine, benazepril, or any other ACE inhibitor
Adverse Reactions	• Constipation • Dizziness • Nausea • Hypotension • Headache • Edema • CHF/pulmonary edema • Fatigue • Dyspnea • Bradycardia • AV block • Rash • Flushing • Elevated hepatic enzymes	• Edema • Dizziness • Flushing • Palpitations	• Cough • Headache • Dizziness • Edema
Drug Interactions	• Other antihypertensive agents (vasodilators, ACE inhibitors, diuretics, and beta-blockers) • Digoxin • Disopyramide • Flecainide • Quinidine • Cimetidine • Lithium • Carbamazepine • Rifampin • Phenobarbital • Cyclosporine • Theophylline • Inhalation anesthetics • Neuromuscular blocking agents	• None known	• Diuretics • Potassium supplements or potassium-sparing diuretics • Lithium

Side tabs: Immunosuppression | Antimicrobials | Cardiovascular Agents | Antiosteoporosis Agents | Antiplatelets | Diabetes Agents | Ulcer Treatment | Diuretics | Other Agents

Continued ...

Calcium Channel Blockers

Brand Name	Calan® SR (verapamil)	Norvasc® (amlopidine)	Lotrel™ (amlopidine and benazepril hydrochloride)
Dosage	• Dose should be individualized by titration • 180 mg/d with upward titrations if adequate response is not obtained	• Dose should be individualized • 5 mg/d with a maximum dose of 10 mg/d	• Dose should be individualized by titration • 2.5 mg to 5 mg amlodipine and 10 mg to 20 mg of benazepril once daily

Editors' Notes:

Calcium channel blockers are considered the drug of first choice for the treatment of posttransplant hypertension since they increase renal blood flow. Nifedipine, isradipine and amlodipine show little interaction with cyclosporine-A. Diltiazem and verapamil elevate cyclosporine-A levels.

Alpha-Adrenergic Receptor Blockers

Immuno-suppression

Antimicrobials

Cardiovascular Agents

Antiosteo-porosis Agents

Antiplatelets

Diabetes Agents

Ulcer Treatment

Diuretics

Other Agents

Brand Name	Hytrin® (terazosin)	Cardura® (doxazosin)	Normodyne® (labetalol) (α- and β-blocker)
Company	Abbott Laboratories	Pfizer	Key
Mechanism of Action	• Block alpha-adrenoreceptors in the brain stem, resulting in reduced sympathetic outflow from the CNS and a decrease in peripheral vascular resistance		
Indication	• Management of hypertension		
Contraindications	• None reported	• Known sensitivity to quinazolines	• Bronchial asthma • Overt cardiac failure • AV block • (>1st degree) • Severe bradycardia
Special Consideration	• Alpha-adrenergic agents can cause marked hypotension after the first dose, especially in the upright position, with syncope and other postural symptoms		
Adverse Reactions	• Dizziness • Headache • Asthenia • Nasal congestion • Peripheral edema • Somnolence • Nausea • Palpitations • Pain-extremities • Paresthesia • Dyspnea • Sinusitis • Back pain • Nervousness • Tachycardia • Blurred vision • Postural hypotension • Impotence	• Dizziness • Nausea • Fatigue • Dyspepsia • Nasal stuffiness • Headache • Ejaculation failure • Dyspnea • Vertigo • Asthenia • Taste distortion • Sexual dysfunction • Edema • Postural hypotension • Rash • Vision abnormality	• Dizziness • Fatigue • Nausea • Vomiting • Dyspepsia • Paresthesia • Nasal stuffiness • Ejaculation failure • Impotence • Edema
Drug Interactions	• None reported	• None reported	• Antidepressants • Beta-blockers • Cimetidine • Nitroglycerin
Dosage	• Dosage must be individualized • Initial dosage is 1 mg at bedtime • Maintenance range is usually 1 mg/d to 5 mg/d	• Dosage must be individualized • Initial dosage is 1 mg given once daily • Depending on standing blood pressure, dose may increased to 2 mg and then if necessary to 4 mg, 8 mg, or 16 mg	• 100 mg twice daily • After 2 or 3 days, dosage may be titrated in increments of 100 mg BID every 2 or 3 days; usual maintenance range is 200 mg to 400 mg twice daily

Immuno-suppression

Antimicrobials

Cardiovascular Agents

Antiosteo-porosis Agents

Antiplatelets

Diabetes Agents

Ulcer Treatment

Diuretics

Other Agents

Central Alpha-Adrenergic Agonist

Brand Name	Catapres® (clonidine)
Company	Boehringer Ingelheim Pharmaceuticals, Inc.
Mechanism of Action	• Stimulation of alpha adrenoreceptors which decrease sympathetic outflow from the CNS
Indication	• Hypertension
Contraindications	• None reported
Special Consideration	• Abrupt withdrawal of clonidine may result in rebound hypertension
Adverse Reactions	• Dry mouth • Drowsiness • Dizziness • Constipation • Sedation
Drug Interactions	• Anti-depressants • CNS depressants: (alcohol, barbiturates, and other sedatives)
Dosage	• Dosage must be individualized • Initial dosage is 0.1 mg twice a day • Maintenance dose is reached with 0.1 mg/d increments until patient reaches the appropriate maintenance range of 0.2 mg/d to 0.6 mg/d given in divided doses

Editors' Notes:

Clonidine is useful in diabetic patients because there is no effect on lipids or on glucose metabolism.

Beta-Blockers

Brand Name	Lopressor® (metoprolol)	Tenormin® (atenolol)
Company	Novartis	AstraZeneca
Mechanism of Action	Not known, but the following have been suggested: • Competitive antagonism of catecholamines at peripheral adrenergic neuron sites, leading to decreased cardiac output • Suppression of renin activity	
Indication	• Management of hypertension	
Contraindications	• Sinus bradycardia • Heart block (>1st degree) • Cardiogenic shock • Overt cardiac failure	
Special Consideration	• Patients with bronchospastic disease should not receive therapy with beta-blockers	
Adverse Reactions	• Tiredness • Dizziness • Depression • Diarrhea • Pruritus/rash • Shortness of breath • Bradycardia • Cold extremities • Arterial insufficiency (Raynaud type) • Palpitations • CHF • Peripheral edema • Hypotension • Bronchospasm • Dyspnea • Nausea • Dry mouth • Gastric pain • Constipation • Flatulence • Heartburn	• Dizziness • Nausea • Bradycardia • Fatigue • Postural hypotension • Vertigo • Diarrhea • Light-headedness • Lethargy
Drug Interactions	• Catecholamine-depleting drugs (reserpine) • Clonidine • Verapamil • Epinephrine (in patients with history of severe anaphylactic reactions to a variety of allergens)	
Dosage	• Dosage must be individualized • Initial dosage is 100 mg/d in single or divided doses alone or added to diuretic therapy • Dosage can be increased at weekly (or longer intervals) until optimal control is achieved, usually at 100 mg/d to 450 mg/d	• Dosage must be individualized • Initial dosage is 50 mg/d given as 1 tablet alone or added to diuretic therapy • If optimal effect is not achieved within 1 to 2 weeks, dosage can be increased to 100 mg/d given as a single tablet

Antimicrobials

Cardiovascular Agents

Antiosteo-porosis Agents

Antiplatelets

Diabetes Agents

Ulcer Treatment

Diuretics

Other Agents

Editors' Notes:

Beta-blockers therapy should be considered in all patients with a significant cardiac history.

Beta-Blockers

Immuno-suppression

Antimicrobials

Cardiovascular Agents

Antiosteo-porosis Agents

Antiplatelets

Diabetes Agents

Ulcer Treatment

Diuretics

Other Agents

Brand Name	Coreg® (carvedilol)
Company	GlaxoSmithKline
Mechanism of Action	Precise mechanism unknown but the following have been suggested • Protection from the cardio-toxic effects of increased catecholamines • Competitive antagonism of catecholamines at peripheral adrenergic neuron sites, leading to decreased cardiac output • Suppression of rennin activity
Indications	• Treatment of mild-to-severe heart failure of ischemic or cardiomyopathic origin to increase survival and to reduce the risk of hospitalization • Management of hypertension • Reduction of cardiovascular mortality in clinically stable patients who have survived the acute phase of a myocardial infarction and have a left ventricular ejection fraction of ≤40%
Contraindications	• Bronchial asthma or related bronchospastic conditions • Heart block (2nd or 3rd degree) • Sick sinus syndrome • Severe bradycardia • Cardiogenic shock • Decompensated heart failure requiring IV inotropic therapy • Hypersensitivity to any components of the product
Warnings	• Abrupt discontinuation of therapy may lead to worsening of angina, MI and ventricular arrhythmia
Special Precautions	• Pregnancy Category C
Adverse Reactions	• AV block • Bradyarrhythmia • Hypotension • Diarrhea • Nausea • Rash • Visual abnormalities • Headache/fatigue
Drug Interactions	• Catecholamine-depleting drugs (reserpine) • Clonidine • Verapermil • Epinephrine (in patients with history of severe anaphylactic reactions to a variety of allergens)
Formulations	• Tablets containing 3.125 mg, 6.25 mg, 12.5 mg, and 25 mg
Dosage	• 3.125-50 mg orally twice daily

ACE Inhibitors

Brand Name	Prinivil® (lisinopril) Zestril® (lisinopril)	Vasotec® (enalapril)	Capoten® (captopril)
Company	Merck & Co., Inc. AstraZeneca	Biovail	Par
Mechanism of Action	• Inhibit angiotensin-converting enzyme (ACE), resulting in decreased plasma angiotensin II which leads to decreased vasopressor activity and decreased aldosterone secretion		
Indication	• Treatment of hypertension, heart failure		
Contraindications	• Hypersensitivity to agent • History of angioedema related to previous ACE inhibitor treatment		
Adverse Reactions	• Dizziness • Headache • Fatigue • Diarrhea • Upper respiratory symptoms • Cough • Nausea • Hypotension • Rash • Orthostatic effects • Asthenia • Chest pain • Vomiting • Dyspnea	• Headache • Dizziness • Fatigue • Diarrhea • Nausea • Rash • Cough • Orthostatic effects • Asthenia • Dyspepsia	• Rash, often with pruritus, and sometimes fever, arthralgia, and eosinophilia • Dysgeusia • Cough • Proteinuria • Hypotension • Tachycardia • Chest pain • Palpitations • Angioedema
Drug Interactions	• Diuretics • Indomethacin • Agents increasing serum potassium • Lithium • Agents causing renin release • Agents affecting sympathetic activity		
Dosage	• Initial dosage is 10 mg/d given once a day • Dose should be individually adjusted according to blood pressure response • Usual maintenance range is 20 mg/d to 40 mg/d in a single dose	• Initial dosage is 5 mg/d given once a day • Dose should be individually adjusted according to blood pressure response • Usual maintenance range is 10 mg/d to 40 mg/d in a single dose or in 2 divided doses	• Dosage must be individualized and captopril tablets should be taken 1 hour before meals • Initial dosage is 25 mg BID or TID • If optimal effect is not achieved within 1 to 2 weeks, dosage can be increased to 50 mg BID or TID

Continued ...

Immuno-suppression

Antimicrobials

Cardiovascular Agents

Antiosteo-porosis Agents

Antiplatelets

Diabetes Agents

Ulcer Treatment

Diuretics

Other Agents

Immuno-suppression

Antimicrobials

Cardiovascular Agents

Antiosteo-porosis Agents

Antiplatelets

Diabetes Agents

Ulcer Treatment

Diuretics

Other Agents

ACE Inhibitors

Additional ACE Inhibitors	Brand	Generic	Company	Dosage Range
	Aceon	Perindopril	Solvay	4 to 8 mg daily
	Accupril	Quinapril	Parke-Davis	10 to 80 mg daily or divided twice daily
	Altace	Ramipril	Monarch	2.5 to 20 mg daily or divided twice daily
	Lotensin	Benazepril	Novartis	5 to 40 mg daily
	Mavik	Trandolapril	Abbott	1 to 8 mg daily
	Monopril	Fosinopril	Bristol-Myers Squibb	5 to 40 mg daily
	Univasc	Moexipril	Schwartz	3.75 to 60 mg daily

Editors' Notes:

Enalapril may cause anemia. It has been used successfully in low doses for posttransplant polycythemia. ACE inhibitors may precipitate renal failure in patients with renal transplant artery stenosis. A rapid rise in serum creatinine after initiation of ACE inhibitor therapy may indicate renal transplant artery stenosis.

Angiotensin II Receptor Antagonists

Immuno-suppression

Antimicrobials

Cardiovascular Agents

Antiosteo-porosis Agents

Antiplatelets

Diabetes Agents

Ulcer Treatment

Diuretics

Other Agents

Brand Name	Cozaar® (losartan)	Diovan® (valsartan)
Company	Merck & Co., Inc.	Novartis
Mechanism of Action	• Selective blockage of binding of angiotensin II to AT_1 receptors	
Indication	• Hypertension	
Contraindications Adverse Reactions	• Hypersensitivity to agent • Fatigue • Abdominal pain • Diarrhea • Nausea • Dyspepsia • Muscle cramps • Myalgia • Pain • Dizziness • Insomnia • Congestion • Cough • Sinus complaints	• Viral infection • Fatigue • Abdominal pain • Headache • Dizziness • Cough • Diarrhea • Edema • Arthralgia • Constipation • Myalgia
Drug Interactions	• Agents increasing serum potassium • Diuretics • Drugs which act on the renin-angiotensin system	
Dosage	• 50 mg once daily • 25 mg in patients who may be volume depleted	• 80 mg once daily • Dosage range from 80 mg to 320 mg

Additional Angiotensin II Antagonists	Brand	Generic	Company	Dosage Range
	Atacand	Candesartan	AstraZeneca	8 to 32 mg daily
	Avapro	Irbesartan	Sanofi-Synthelabo	150 mg to 300 mg daily
	Micardis	Telmisartan	Boehringer Ingelheim	20 to 80 mg daily
	Teveten	Eprosartan	Unimed	400 to 800 mg daily
	Renicar	Olmesartan Medoxomil	Sankyo Pharmaceuticals	20 to 40 mg daily

Editors' Notes:

These agents have toxicities similar to ACE inhibitors but less cough. Anemia has been noted in patients on calcineurin inhibitors.

Cholesterol-Lowering Agents

Brand Name	Mevacor® (lovastatin)	Pravachol® (pravastatin)
Company	Merck & Co., Inc.	Bristol-Myers Squibb Co.
Mechanism of Action	• Inhibit HMG-CoA reductase, preventing the conversion of HMG-CoA to mevalonate, an early step in cholesterol biosynthesis • Depletion of cellular cholesterol stimulates the production of cell surface receptors that recognize LDL, leading to increased catabolism of LDL cholesterol	
Indication	• Adjunct to diet for the reduction of elevated total and LDL cholesterol levels in patients with primary hypercholesterolemia (Types IIa and IIb) when the response to a diet restricted in saturated fat and cholesterol and to other nonpharmacologic measures has been inadequate	
Contraindications	• Hypersensitivity to any component of these medications • Active liver disease • Pregnancy and lactation (Pregnancy Category X)	
Warnings	• Liver function tests should be performed prior to treatment, at 6 and 12 weeks after initiation of therapy and periodically thereafter. • Rhabdomyolysis with acute renal failure has been reported with related drugs in this class.	
Adverse Reactions	• Headache • Flatus • Abdominal pain and cramps • Diarrhea • Rash/Pruritus • Constipation • Nausea • Dyspepsia • Myalgia • Dizziness • Heartburn • Blurred vision • Muscle cramps	• Localized pain • Nausea/vomiting • Common cold • Diarrhea • Headache • Abdominal pain • Cardiac chest pain • Constipation • Chest pain • Flatulence • Dizziness • Heartburn • Myalgia • Influenza • Urinary abnormality
Drug Interactions	• Cyclosporine • Gemfibrozil • Niacin • Erythromycin	• Cholestyramine/Colestipol • Warfarin • Cimetidine • Digoxin
Formulations	• Tablets containing 10 mg, 20 mg, and 40 mg • Extended-release tablets containing 10 mg, 20 mg, 40 mg, and 60 mg	• Tablets containing 10 mg, 20 mg, 40 mg, and 80 mg
Dosage	• Dosage should be individualized • Initial dosage is 20 mg/d taken once daily with the evening meal • Patients on immunosuppressive therapy should be started on 10 mg/d and should not exceed 20 mg/d • Maximum recommended dose is 80 mg/d	• Dosage should be individualized • Starting dose is 10 mg/d or 20 mg/d taken once daily at bedtime • Patients on immunosuppressive therapy should be started on 10 mg/d and should not exceed 20 mg/d

Cholesterol-Lowering Agents

Brand Name	Lescol® (fluvastatin sodium)	Zocor® (simvastatin)	Lipitor® (atorvastatin)
Company	Novartis	Merck & Co., Inc.	Pfizer
Mechanism of Action	• Inhibit HMG-CoA reductase, preventing the conversion of HMG-CoA to mevalonate, an early step in cholesterol biosynthesis • Depletion of cellular cholesterol stimulates the production of cell surface receptors that recognize LDL, leading to increased catabolism of LDL cholesterol		
Indication	• Adjunct to diet for the reduction of elevated total cholesterol, LDL, apo B, and TG levels in patients with primary hypercholesterolemia (heterozygous familial and nonfamilial), mixed dyslipidemia (Fredrickson types IIa and IIb), elevated TG (type IV) and primary dysbetalipoproteinemia (type III) • Adjunct to other lipid lowering treatments for homozygous familial hypercholesterolemia		
Contraindications	• Hypersensitivity to any component of these medications • Active liver disease • Pregnancy and lactation (Pregnancy Category X)		
Warnings	• Liver function tests should be performed prior to treatment, at 12 weeks after initiation of therapy and periodically thereafter. • Rhabdomyolysis with acute renal failure has been reported with related drugs in this class.		
Adverse Reactions	Adverse events with a higher incidence in fluvastatin-treated patients than placebo patients • Arthropathy • Exercise-related muscle pain • Sinusitis • Bronchitis • Dyspepsia • Diarrhea • Abdominal pain • Nausea • Insomnia • Fatigue	• Abdominal pain • Asthenia • Constipation • Diarrhea • Dyspepsia • Flatulence • Nausea • Headache • Upper respiratory tract infection	• Headache • Constipation • Flatulence • Dyspepsia • Abdominal pain • Arthralgia • Myalgia • Bronchitis • Pruritus • Rhabdomyolysis • Increases in serum transaminases and CPK
Drug Interactions	• Gemfibrozil • Niacin • Erythromycin • Cholestyramine • Digoxin • Cimetidine/ranitidine/omeprazole • Rifampicin • Warfarin	• Itraconazole • Gemfibrozil • Niacin • Erythromycin • Propranolol • Digoxin • Warfarin	• Antacids • Colestipol • Digoxin • Erythromycin • Oral contraceptives • Fibrates • Niacin • Azole antifungals
Formulation	• Capsules containing 20 mg and 40 mg	• Tablets containing 5 mg, 10 mg, 20 mg, 40 mg, and 80 mg	• Tablets containing 10 mg, 20 mg, 40 mg, and 80 mg

Continued ...

Immuno-suppression

Antimicrobials

Cardiovascular Agents

Antiosteo-porosis Agents

Antiplatelets

Diabetes Agents

Ulcer Treatment

Diuretics

Other Agents

Immuno-suppression

Antimicrobials

Cardiovascular Agents

Antiosteo-porosis Agents

Antiplatelets

Diabetes Agents

Ulcer Treatment

Diuretics

Other Agents

Cholesterol-Lowering Agents

Brand Name	Lescol® (fluvastatin)	Zocor® (simvastatin)	Lipitor (atorvastatin)
Dosage	• Initial dosage is 20 mg/d at bedtime • Recommended dosage range is 20 mg/d to 40 mg/d as a single dose	• Dosage should be individualized according to baseline LDL-C levels • Starting dosage range is 5 mg/d to 10 mg/d • Maximum dosage is 40 mg/d	• Initial doses of 10 mg once daily, individualized according to LDL concentrations, to a maximum of 80 mg once daily

Editors' Notes:

Cyclosporine prolongs the half-life of statin drugs. Therefore, in transplant recipients, the preferred dose of these cholesterol-lowering agents is half of the recommended dose. Some of these agents have been reported to cause rhabdomyolysis when given to patients receiving cyclosporine-A.

Cholesterol-Lowering Agents

Brand Name	Crestor® (rosuvastatin)
Company	AstraZeneca
Mechanism of Action	• Inhibits HMG-CoA reductase and prevents the conversion of HMG-CoA to mevalonate, an early step in cholesterol synthesis • Depletion of cellular cholesterol stimulates the production of cell surface receptors that recognize LDL, leading to increased catabolism of LDL cholesterol.
Indications	• Adjunct to diet for the reduction of elevated total-C, LDL-C, and ApoB, nonHDL-C, and TG levels and to increase HDL-C in patients with primary hypercholesterolemia and mixed dyslipidemia • Adjunct to diet for the treatment of patients with elevated serum TG levels
Contraindications	• Hypersensitivity to rosuvastatin or any components of the product · Active liver disease or persistently elevated liver enzymes
Warnings	• Liver function tests should be performed prior to treatment, at 12 weeks after initiation of therapy and periodically thereafter • Rhabdomyolysis with acute renal failure has been reported
Special Precautions	• Pregnancy Category X
Adverse Reactions	• Diarrhea/constipation • Nausea • Drug-induced myopathy • Rhabdomyolysis • Nephrotoxicity
Drug Interactions	• Gemfibrozil • Cyclosporine • Fluconazole • Niacin • Warfarin
Formulations	• Tablets containing 5 mg, 10 mg, 20 mg, and 40 mg
Dosage	• 5-40 mg orally once daily

Immuno-suppression | Antimicrobials | Cardiovascular Agents | Antiosteoporosis Agents | Antiplatelets | Diabetes Agents | Ulcer Treatment | Diuretics | Other Agents

Cholesterol-Lowering Agents

Immuno-suppression

Antimicrobials

Cardiovascular Agents

Antiosteo-porosis Agents

Antiplatelets

Diabetes Agents

Ulcer Treatment

Diuretics

Other Agents

Brand Name	**WelChol® (colesevelam hydrochloride)**
Company	Sankyo
Class	• Bile Acid Sequestrant
Mechanism of Action	• Binds bile acids in the intestine and preventing their absorption
Indications	• Adjunct therapy with diet and exercise for the reduction of elevated LDL cholesterol in patients with primary hypercholesterolemia
Contraindications	• Known hypersensitivity to colesevelam hydrochloride • Bowel obstruction
Special Precautions	• Biliary obstruction/atresia • Dysphagia/swallowing disorders • Gastrointestinal disorders/bowel disease • Primary biliary cirrhosis • Triglyceride level >300 mg/dL • Vitamin K or other fat soluble vitamin deficiencies • Pregnancy Category B
Adverse Reactions	• Constipation • Dyspepsia • Myalgia • Asthenia
Drug Interactions	• Diltiazem • Mycophenolate sodium • Mycophenolate mofetil • Ezetimibe • Fenofibrate
Formulations	• 625 mg tablets
Dosage	• 1875 mg orally twice a day with meals

Other Bile-Acid Sequestrant	*Brand*	*Generic*	*Company*	*Dosage Range*
	Questran ®	Cholestyramine	Par	4 to 8 grams once or twice daily
	Colestid®	Colestipol	Pharmacia	Tablets: 2 to 16 grams orally daily or in divided doses Powder: 5 to 30 grams orally daily or in divided doses

Cholesterol-Lowering Agents

Brand Name	Zetia® (ezetimibe)
Company	Merck/Schering Plough
Class	• Selective cholesterol absorption inhibitor
Mechanism of Action	• Inhibits passage of dietary and biliary cholesterol across the brush border of the small intestine, with minimal or no effect on absorption of other soluble food nutrients
Indications	• Adjunct to diet for the reduction of elevated total-C, LDL-C, and Apo B in patients with primary hypercholesterolemia • In combination with an HMG-CoA reductase inhibitor, as adjunctive therapy to diet for the reduction of elevated total-C, LDL-C, and Apo B in patients with primary hypercholesterolemia
Contraindications	• Hypersensitivity to ezetimibe · Active liver disease or persistently elevated liver enzymes (when coadministered with a statin)
Special Precautions	• Pregnancy Category C
Adverse Reactions	• Diarrhea • Thrombocytopenia • Elevated liver function tests • Hypersensitivity/anaphylaxis reactions • Angioedma • Headache • Rash/pruritus • Sinusitis
Drug Interactions	• Cholestyramine • Cyclosporine • Fenofibrate • Gemfibrozil • Colestipol • Colesevelam
Formulations	• 10 mg tablets
Dosage	• 10 mg orally once daily

Immuno-suppression

Antimicrobials

Cardiovascular Agents

Antiosteo-porosis Agents

Antiplatelets

Diabetes Agents

Ulcer Treatment

Diuretics

Other Agents

Chapter 4
Antiosteoporosis Agents

Antiosteoporosis Agents Overview

- Alendronate Sodium

- Risedronate

- Ibandronate Sodium

- Etidronate Disodium

- Calcitonin-Salmon

Antiosteoporosis Agents Overview

Agent	Dosage
Alendronate Sodium	*Postmenopausal Osteoporosis* • 10 mg/d with a full glass of *plain* water, or 70 mg once weekly *Paget's Disease* • 40 mg/d for 6 months
Risedronate	*Treatment and Prevention of Postmenopausal Osteoporosis or Glucocorticoid-Induced Osteoporosis* • 5 mg orally daily or 35 mg weekly *Paget's Disease* • 30 mg orally once daily for 2 months
Ibandronate Sodium	• 2.5 mg orally daily or 150 mg orally once monthly or • 3 mg IV every 3 months
Etidronate Disodium	*Paget's Disease* • 5 mg/kg/d to 10 mg/kg/d for a period not to exceed 6 months or 11 mg/kg/d to 20 mg/kg/d for no longer than 3 months *Heterotopic Ossification* • Total hip replacement: 20 mg/kg/d for 1 month before surgery and 3 months after surgery • Spinal cord injury: 20 mg/kg/d for 2 weeks followed by 10 mg/kg/d for 10 weeks
Calcitonin-Salmon	• 200 IU/d intranasally, alternating nostrils daily

Editors' Notes:

Alendronate sodium is indicated for the prevention of glucocorticoid-induced bone loss. Rapid bone loss occurs after transplantation, placing these patients at increased risk for fractures. Because of improving long-term survival, attention to metabolic bone disease should occur early in the posttransplant course so that the risk of complications might be lessened.

Immunosuppression | Antimicrobials | Cardiovascular Agents | Antiosteoporosis Agents | Antiplatelets | Diabetes Agents | Ulcer Treatment | Diuretics | Other Agents

Alendronate Sodium

Immuno-suppression

Antimicrobials

Cardiovascular Agents

Antiosteo-porosis Agents

Antiplatelets

Diabetes Agents

Ulcer Treatment

Diuretics

Other Agents

Brand Name	**Fosamax®**
Company	Merck & Co., Inc.
Class	• Hormone-free biphoshanate antiosteoporosis agent
Mechanism of Action	• Inhibits osteoclast-mediated bone resorption, thereby reducing the rate of bone loss and increasing bone density
Indications	• Postmenopausal osteoporosis—prevention and treatment • Paget's disease having alkaline phosphatase at least two times the upper limit of normal, or in patients who are symptomatic or at risk for further complications • Treatment ot increase bone mass in men with osteoporosis • Treatment of glucocorticoid induced osteoporosis
Contraindications	• Hypersensitivity to any component of this product • Hypocalcemia • Abnormalities of the esophagus which delay emptying • Inability to stand or sit upright for at least 30 minutes
Special Precautions	• Not recommended for patients with renal insufficiencies • Should be used with caution in patients with gastrointestinal problems • Hypocalcemia should be corrected before initiating therapy • Pregnancy Category C
Adverse Reactions	• Abdominal pain • Nausea • Dyspepsia • Constipation • Diarrhea • Flatulence • Acid regurgitation • Esophageal ulcer • Vomiting • Dysphagia • Abdominal distention • Gastritis • Taste perversion • Headache • Dizziness • Muscle cramp • Musculoskeletal pain
Drug Interactions	• Ranitidine • Aspirin • Calcium supplements/antacids • NSAIDs
Formulation	• Tablets containing 5 mg, 10 mg, 35 mg, 40 mg, 70 mg • Oral solution 70 mg/75 mL
Dosage	*Treatment of Postmenopausal Osteoporosis or to Increase Bone Mass in Men with Osteoporosis* • 10 mg/d or 70 mg once weekly *Paget's Disease* • 40 mg/d for 6 months *Prevent Postmenopausal Osteoporosis or Treatment of Glucocoricoid Induced Osteoporosis* • 5 mg daily or 35 mg weekly

Editors' Notes:

Major side effects include GI symptoms. Renal insufficiency may preclude use. Take with a full glass of water, 30 minutes prior to first food, beverage or medication. Remain upright for at least 30 minutes after taking.

Risedronate

Brand Name	**Actonel®**
Company	Procter & Gamble Pharmaceuticals
Class	• Hormone-free bisphosphanate antiosteoporosis agent
Mechanism of Action	• Inhibits osteoclast-mediated bone resorption and modulates bone metabolism • Reduces rate of bone loss and increases bone density
Indications	• Prevention and treatment of postmenopausal osteoporosis • Treatment of glucocorticoid-induced osteoporosis • Induction of remission for Paget's disease in patients with serum alkaline phosphatase that is at least 2 times the upper limit of normal or who are symptomatic, or who are at risk for future complications
Contraindications	• Known hypersensitivity to risedronate or to any components of the product • Hypocalcemia • Inability to stand or sit upright for at least 30 minutes
Warnings	• May cause upper gastrointestinal disorders such as dysphagia, esophagitis, and esophageal or gastric ulcer
Special Precautions	• Cancer patients undergoing dental procedures are at an increased risk of osteonecrosis of the jaw • Treatment is not recommended in patients with severe renal impairment (creatinine clearance <30 mL/min) • Hypocalcemia and other disturbances of bone and mineral metabolism should be corrected prior to initlating therapy • Pregnancy Category C

Adverse Reactions	• Chest pain
	• Hypertension
	• Peripheral edema
	• Colitis
	• Rash/pruritus
	• Gastrointestinal hemorrhage
	• Diarrhea/constipation
	• Nausea/vomiting/dyspepsia
	• Flatulence
	• Gastroenteritis
	• Xerostomia
	• Hypersensitivity reaction
	• Arthralgia/arthritis
	• Back pain/neck pain
	• Aseptic necrosis of bone (primarily of the jaw)
	• Bone pain
	• Myalgia
	• Asthenia
	  • Headache
	  • Dizziness
	  • Vertigo
	  • Dry eye
	  • Cataract
	  • Anxiety
	  • Depression
	  • Nephrolithiasis
	  • Urinary tract infection
	  • Benign prostatic hyperplasia
	  • Cough
	  • Bronchitis
	  • Pharyngitis
	  • Rhinitis
	  • Sinusitis
	  • Pneumonia
	  • Influenza-like illness
	  • Accidental injury
	  • Pain

Drug Interactions	• Calcium supplements/magnesium supplements • Aspirin • Non-Steroidal Anti-Inflammatory Drugs
Formulations	• 5 mg film-coated tablets • 30 mg film-coated tablets • 35 mg film-coated tablets
Dosage	*Treatment and Prevention of Postmenopausal Osteoporosis or Glucocorticoid-Induced Osteoporosis* • 5 mg orally daily or 35 mg weekly *Paget's Disease* • 30 mg orally once daily for 2 months

Immuno-suppression

Antimicrobials

Cardiovascular Agents

Antiosteo-porosis Agents

Antiplatelets

Diabetes Agents

Ulcer Treatment

Diuretics

Other Agents

Ibandronate Sodium

Immuno-suppression

Antimicrobials

Cardiovascular Agents

Antiosteo-porosis Agents

Antiplatelets

Diabetes Agents

Ulcer Treatment

Diuretics

Other Agents

Brand Name	**Boniva®**
Company	Roche Laboratories
Class	• Hormone-free bisphosphanate antiosteoporosis agent
Mechanism of Action	• Inhibits osteoclast-mediated bone resorption and modulates bone metabolism • Reduces rate of bone loss and increases bone density
Indications	• Prevention and treatment of postmenopausal osteoporosis
Contraindications	• Known hypersensitivity to ibandronate or to any components of the product • Uncorrected hypocalcemia • Inability to stand or sit upright for at least 30 minutes
Warnings	• May cause upper gastrointestinal disorders such as dysphagia, esophagitis, and esophageal or gastric ulcer
Special Precautions	• Cancer patients undergoing dental procedures are at an increased risk of osteonecrosis (primarily of the jaw) • Treatment is not recommended in patients with severe renal impairment (creatinine clearance <30 mL/min) • Hypocalcemia and other disturbances of bone and mineral metabolism should be corrected prior to initiating therapy • Severe and occasionally incapacitating bone, joint and/or muscle pain has been reported • Pregnancy Category C
Adverse Reactions	• Hypertension • Arthralgia/arthropathy • Rash • Back pain • Hypercalciuria • Aseptic necrosis of bone • Hypocalcemia (primarily of the jaw) • Hypomagnesemia • Bone pain • Hypophosphatemia • Myalgia • Hypercholesterolemia • Headache • Esophagitis • Dizziness • Abdominal pain • Bronchitis • Diarrhea/constipation • Pharyngitis • Nausea/vomiting/dyspepsia • Pneumonia • Thrombocytopenia • Urinary tract infection • Hepatotoxicity • Influenza-like illness • Hypersensitivity reaction • Fever
Drug Interactions	• Calcium supplements/magnesium supplements • Aspirin • Non-steroidal anti-inflammatory drugs
Formulations	• 2.5 mg tablets • 150 mg tablets • 1 mg/mL in 5 mL single-use prefilled syringe
Dosage	• 2.5 mg orally daily or 150 mg orally once monthly or • 3 mg IV every 3 months

Etidronate Disodium

Brand Name	Didronel®
Company	Procter & Gamble Pharmaceuticals, Inc.
Class	• Bone metabolism regulator
Mechanism of Action	• Inhibits formation, growth, and dissolution of hydroxyapatite crystals and their amorphous precursors by chemisorption to calcium phosphate surfaces
Indications	• Symptomatic Paget's disease of bone • Heterotopic ossification
Contraindications	• Known hypersensitivity to Etidronate disodium • Clinically overt osteomalacia
Warning	• In Paget's patients, onset of effect may be slow
Special Precautions	• Patients should maintain adequate intake of calcium and vitamin D • Patients with enterocolitis may experience diarrhea • Etidronate is not metabolized and is excreted intact via the kidney • Pregnancy Category C
Adverse Reactions	• Gastrointestinal complaints • Increased pain at previously asymptomatic pagetic sites • Alopecia • Arthropathies • Bone fracture • Esophagitis/glossitis • Hypersensitivity reactions
Drug Interactions	• None reported
Formulation	• 200 mg tablets
Dosage	*Paget's Disease* • 5 mg/kg/d to 10 mg/kg/d for a period not to exceed 6 months or 11 mg/kg/d to 20 mg/kg/d for no longer than 3 months *Heterotopic Ossification* • Total hip replacement: 20 mg/kg/d for 1 month before surgery and 3 months after surgery • Spinal cord injury: 20 mg/kg/d for 2 weeks followed by 10 mg/kg/d for 10 weeks

Immuno-suppression

Antimicrobials

Cardiovascular Agents

Antiosteo-porosis Agents

Antiplatelets

Diabetes Agents

Ulcer Treatment

Diuretics

Other Agents

Calcitonin-Salmon

Brand Name	**Miacalcin®**
Company	Novartis
Class	• Synthetic polypeptide hormone with the same linear sequence of amino acids as calcitonin of salmon origin
Mechanism of Action	• Causes transient inhibition of the bone resorptive process possibly through inhibition of osteoclast function with loss of the ruffled osteoclast border responsible for resorption of bone
Indication	• Postmenopausal osteoporosis
Contraindication	• Clinical allergy to Calcitonin-salmon
Warning	• Possibility of systemic allergic reaction exists because Calcitonin-salmon is a polypeptide
Special Precautions	• Periodic nasal examinations are recommended • Pregnancy Category C
Adverse Reactions	• Rhinitis and other nasal symptoms • Back pain • Arthralgia • Epistaxis • Headache
Drug Interactions	• No drug interaction studies have been conducted • No drug interactions have been observed
Formulation	• Metered dose solution in 2-mL fill glass bottles with a dose strength of 200 IU per activation (0.09 mL/puff)
Dosage	• 200 IU/d intranasally, alternating nostrils daily

Editors' Notes:

New agent with little experience in transplant recipients.

Chapter 5
Antiplatelets

Antiplatelets Overview

- Aspirin
- Clopidogrel

Antiplatelets Overview

Agent	Dosage
Aspirin	*Transient Ischemic Attack* • 1300 mg/d given in divided doses of 650 mg 2 times a day or 325 mg 4 times a day *Myocardial Infarction* • 325 mg/d
Clopidogrel	*Recent MI, recent stroke, or established peripheral arterial disease* • 75 mg orally once daily. *Acute Coronary Syndrome* • Loading dose — 300 mg orally • Maintenance dose — 75 mg once daily

Immuno-suppression

Antimicrobials

Cardiovascular Agents

Antiosteo-porosis Agents

Antiplatelets

Diabetes Agents

Ulcer Treatment

Diuretics

Other Agents

Aspirin

Immuno-suppression
Antimicrobials
Cardiovascular Agents
Antiosteo-porosis Agents
Antiplatelets
Diabetes Agents
Ulcer Treatment
Diuretics
Other Agents

Names of Commonly Used Brands	• Anacin® • Easprin® • Ascriptin® • Ecotrin® • Bayer® Aspirin • Norwich® Aspirin • Bufferin®
Mechanism of Action	• Inhibits the enzyme prostaglandin cyclooxygenase in platelets, which prevents the formation of the aggregating agent thromboxane A_2 • May also inhibit formation of prostacyclin, a platelet aggregation inhibitor
Indication	• Antiplatelet prophylaxis to reduce risk of recurrent transient ischemic attacks (TIA) and myocardial infarction (MI)
Contraindication	• Hypersensitivity to aspirin
Adverse Reactions	• Stomach pain • Heartburn • Nausea/vomiting • Increased rates of gross gastrointestinal bleeding • Small increases in systolic blood pressure • Slight increases in BUN and uric acid levels
Drug Interactions	• Acetaminophen • Acidifiers, urinary (ammonium chloride, ascorbic acid, potassium or sodium phosphates) • Adrenocorticoids, glucocorticoids, and corticotropin • NSAIDs • Alkalizers, urinary (carbonic anhydrase inhibitors, citrates, sodium bicarbonate, antacids) • Anticoagulants • Heparin or thrombolytic agents • Hydantoin • Antidiabetic agents • Cefamandole, cefoperazone, cefotetan, moxalactam, plicamycin, valproic acid • Furosemide • Laxatives, cellulose-containing • Methotrexate • Nifedipine, verapamil • Vancomycin • Platelet aggregation inhibitors • Probenecid, sulfinpyrazone • Salicylic acid • Zidovudine
Dosage	*TIA* • 1300 mg/d given in divided doses of 650 mg 2 times a day or 325 mg 4 times a day *MI* • 325 mg/d

Editors' Notes

Our usual practice is to use 81 mg/day for cardiovascular prophylaxis.

Clopidogrel

Immuno-suppression

Antimicrobials

Cardiovascular Agents

Antiosteo-porosis Agents

Antiplatelets

Diabetes Agents

Ulcer Treatment

Diuretics

Other Agents

Brand Name	Plavix®
Company	Sanofi-Aventis
Class	• Inhibitor of ADP-induced platelet aggregation
Mechanism of Action	• Selective, irreversible inhibition of adenosine diphosphate-induced platelet aggregation • Complete recovery of platelet function did not occur until 7 days after ingestion of the last dose of clopidogrel
Indications	• To reduce the rate of a combined endpoint of new ischemic stroke, new myocardial infarction (MI), and other vascular death in patients with recent MI, recent stroke or established peripheral arterial disease • To decrease the rate of a combined endpoint of cardiovascular death, MI, stroke, or refractory ischemia in patients with acute coronary syndrome
Contraindications	• Hypersensitivity to clopidogrel or to any components of the product • Active pathologic bleeding, such as peptic ulcers or intracranial hemorrhage
Warnings	• Rare cases of thrombotic thrombocytopenic purpura have been reported
Special Precautions	• Use with caution in patients with severe hepatic disease or severe renal impairment • Pregnancy Category B
Adverse Reactions	• Edema • Hypertension • Rash/pruritus • Hypercholesterolemia • Gastrointestinal hemorrhage • Diarrhea/nausea/abdominal pain/dyspepsia • Anemia • Purpura • Epistaxis • Increased risk for post-operative hemorrhage • Thrombotic thrombocytopenic purpura • Elevated liver function tests • Allergic reactions (rare) • Arthralgia/back pain • Depression • Headache • Dizziness • Intracranial hemorrhage • Fatigue
Drug Interactions	*Increased risk for bleeding when used with any of the following:* • Glycoprotein IIb/IIIa Inhibitors • Heparin • Alteplase • Low-molecular weight heparin • Aspirin • Warfarin • Non-steroidal anti-inflammatory drugs
Formulations	• 75 mg tablets
Dosage	*Recent MI, recent stroke, or established peripheral arterial disease* • 75 mg orally once daily. *Acute Coronary Syndrome* • Loading dose—300 mg orally • Maintenance dose—75 mg once daily

Chapter 6
Diabetes Agents

Diabetes Agents Overview

- Insulin

- Glyburide

- Glipizide

- Metformin Hydrochloride

- Pioglitazone

- Rosaglitazone

Diabetes Agents Overview

Agent	Dosage
Insulin	*Insulin, Aspart Solution*
	Insulin, Glargine Solution
	Insulin, Human Isophane Suspension
	Insulin, Human NPH
	Insulin, Human Regular
	Insulin, Human Regular and Human NPH Mixture
	Insulin, Human, Zinc Suspension
	Insulin, Lispro Solution
	Insulin, NPH
	Insulin, Regular
	Insulin, Regular and NPH Mixture
	Insulin, Zinc Crystals
	Insulin, Zinc Suspension
	Insulin, Regular Concentrate
	Insulin, Detemir
	Insulin, Glulisine
	Insulin, Human (inhalation)
Glyburide	*Diaβeta®* • Usual starting dose range is 2.5 mg/d to 5 mg/d administered with breakfast or the first main meal • Usual maintenance range is 1.25 mg/d to 20 mg/d given as a single dose or in divided doses • Increments should be no greater than 2.5 mg at weekly intervals *Micronase®* • Usual starting dose range is 2.5 mg/d to 5 mg/d administered with breakfast or the first main meal • Usual maintenance range is 1.25 mg/d to 20 mg/d given as a single dose or in divided doses • Increments should be no greater than 2.5 mg at weekly intervals *Glynase®* • Patients should be retitrated when transferred from other oral hypoglycemic agents • Usual starting dose range is 1.5 mg/d to 3 mg/d administered with breakfast or the first main meal • Usual maintenance range is 0.75 mg/d to 12 mg/d given as a single dose or in divided doses • Increments should be no greater than 1.5 mg at weekly intervals

Continued ...

Immuno-suppression

Antimicrobials

Cardiovascular Agents

Antiosteo-porosis Agents

Antiplatelets

Diabetes Agents

Ulcer Treatment

Diuretics

Other Agents

Immuno-suppression		

Diabetes Agents Overview

Glipizide	*Glucotrol®* • Recommended initial dose is 5 mg/d given before breakfast • Dosage adjustments should be made in increments of 2.5 mg to 5 mg • Maximum once-daily dose is 15 mg; dosages above 15 mg should be divided, given before meals of adequate caloric content, and should not exceed 40 mg/d
Metformin **Hydrochloride**	• Recommended starting dosage is 500 mg twice daily or 850 mg daily with meals • Dosage should be increased by one 500 mg tablet each week, or by one 850 mg tablet every 2 weeks • Individualize dosage based on effectiveness and tolerance, up to a maximum of 2550 mg/day given in divided doses
Pioglitazone	• Actos® Initial dose: 15 mg or 30 mg daily; maximum 45 mg daily
Rosaglitazone	• Avandia® Initial dose: 4 mg daily or in divided doses twice daily; maximum 8 mg daily or in divided doses twice daily

Side tabs: Immuno-suppression · Antimicrobials · Cardiovascular Agents · Antiosteo-porosis Agents · Antiplatelets · Diabetes Agents · Ulcer Treatment · Diuretics · Other Agents

Insulin

Isophane insulin suspension (NPH) (insulin combined with protamine and zinc)

Brand Name	Company	Ingredient	Concentration	Product
NPH-N	Novo Nordisk	purified pork	100 u/mL	10 mL vials
Pork NPH Iletin II	Lilly	purified pork	100 u/mL	10 mL bottles
Humulin N	Lilly	human insulin (rDNA)	100 u/mL	10 mL bottles
Novolin N	Novo Nordisk	human insulin (rDNA)	100 u/mL	10 mL vials
Novolin N Penfill	Novo Nordisk	human insulin (rDNA)	100 u/mL	1.5 mL cartridge
Novolin N Prefilled	Novo Nordisk	human insulin (rDNA)	100 u/mL	1.5 mL prefilled syringes

Isophane insulin suspension and regular insulin injection (combination)

Brand Name	Company	Ingredient	Concentration	Product
Humulin 70/30	Lilly	70% isophane 30% regular human	100 u/mL	10 mL bottles
Novolin 70/30	Novo Nordisk	70% isophane 30% regular human	100 u/mL	10 mL vials
Novolin 70/30 Penfill	Novo Nordisk	70% isophane 30% regular human	100 u/mL	1.5 mL cartridge
Humulin 50/50	Lilly	50% isophane 50% regular human	100u/mL	10 mL vials
Novolin 70/30 Prefilled	Novo Nordisk	70% isophane 30% regular human	100 u/mL	1.5 mL prefilled syringes

Insulin zinc suspension (Lente) (70% crystalline and 30% amorphous)

Brand Name	Company	Ingredient	Concentration	Product
Lente L	Novo Nordisk	purified pork	100 u/mL	10 mL vials
Humulin L	Lilly	human insulin (rDNA)	100 u/mL	10 mL bottles
Novolin L	Novo Nordisk	human insulin (rDNA)	100 u/mL	10 mL vials

Insulin zinc suspension, extended (Ultralente)

Brand Name	Company	Ingredient	Concentration	Product
Humulin U Ultralente	Lilly	human insulin (rDNA)	100 u/mL	10 mL bottles

Insulin glargine solution

Brand Name	Company	Ingredient	Concentration	Product
Lantus	Aventis	human insulin (rDNA)	100 u/mL	5 mL vial 10 mL vial 3 mL cartridges

Continued ...

Insulin

Insulin detemir injection

Brand Name	Company	Ingredient	Concentration	Product
Levemir®	Novo Nordisk	insulin detemir (rDNA)	100 u/mL	3 mL penfill cartridges 3 mL prefilled syringes 10 mL vials

Insulin glulisine injection

Brand Name	Company	Ingredient	Concentration	Product
Apidra®	Aventis	insulin glulisine (rDNA)	100 u/mL	3 mL cartridges for *OptiClik* 10 mL vials

Insulin lispro injection

Brand Name	Company	Ingredient	Concentration	Product
Humalog®	Lilly	insulin lispro (rDNA)	100 u/mL	1.5 mL and 3 mL cartridges 10 mL vials

Insulin aspart injection

Brand Name	Company	Ingredient	Concentration	Product
Novolog®	Novo Nordisk	insulin aspart (rDNA)	100 u/mL	3 mL Penfill cartridges 3 mL Flexpen syringes 10 mL vials

Regular Insulin

Brand Name	Company	Ingredient	Concentration	Product
Regular Purified Pork	Novo Nordisk	purified pork	100 u/mL	10 mL bottles
Humulin R	Lilly	human insulin (rDNA)	100 u/mL	10 mL bottles
Novolin R	Novo Nordisk	human insulin (rDNA)	100 u/mL	10 mL vials
Velosulin Human BR	Novo Nordisk	semisynthetic human insulin	100 u/mL	10 mL vials
Novolin R Penfill	Novo Nordisk	human insulin (rDNA)	100 u/mL	1.5 mL cartridge use with Novo Pen, Novolin Pen
Novolin R Prefilled	Novo Nordisk	human insulin (rDNA)	100 u/mL	1.5 mL prefilled syringe
Regular Concentrated	Lilly	human insulin	500 u/mL	20 mL vials

Continued ..

Insulin

Immuno-suppression

Antimicrobials

Cardiovascular Agents

Antiosteo-porosis Agents

Antiplatelets

Diabetes Agents

Ulcer Treatment

Diuretics

Other Agents

Insulin human (inhalation)

Brand Name	Company	Ingredient	Concentration	Product
Exubera®	Pfizer	Human insulin (rDNA)	None	1 mg powder for inhalation 2 mg powder for inhalation

Subcutaneous	Onset (hrs)	Peak (hrs)	Duration (hrs)	Compatible Mixed With
Rapid acting				
Insulin Regular (Regular)	0.5-1	2.5-5	8-12	All
Prompt Insulin Zinc Suspension (Semilente)	1-1.5	5-10	12-16	Lente
Lispro Insulin Solution (Humalog)	0.25	0.5-1.5	6-8	Ultralente, NPH
Aspart Insulin Solution (Novolog)	0.25	1-3	3-5	NPH
Insulin glulisine	0.3	1	5-6	NPH
Insulin human (inhaled)	0.2-0.3	2	6	None
Intermediate acting				
Isophane Insulin Suspension (NPH)	1-1.5	4-12	24	Regular
Insulin Zinc Suspension (Lente)	1-2.5	7-15	24	Regular, Semilente
70% Isophane Insulin and 30% Regular Insulin (70/30)	0.5-1	4-8	24	
Long acting				
Protamaine Zinc Insulin Suspension (PZI)	4-8	14-24	36	Regular
Extended Insulin Zinc Suspension (Ultralente)	4-8	10-30	>36	Regular, Semilente
Insulin Glargine Solution (Lantus)	1	None	24	None
Insulin detemir	2-4	6-8	6-24	None

Mechanism of Action	• Controls the storage and metabolism of carbohydrates, proteins, and fats, influencing cellular growth, enzyme activation and inhibition, and alterations in protein and fat metabolism

Continued ...

Immuno-suppression

Insulin

Indications	• Diabetes mellitus, insulin-dependent • Diabetes mellitus, non-insulin-dependent • Hyperkalemia
Contraindication	• None reported
Adverse Reactions	• Hypoglycemia (anxious feeling, cold sweats, confusion, cool, pale skin, difficulty in concentration, drowsiness, excessive hunger, headache, nausea, nervousness, rapid pulse, shakiness, unusual tiredness or weakness, vision changes)
Drug Interactions	• Adrenocorticoids, glucocorticoids • Amphetamines • Baclofen • Contraceptives, oral, estrogen-containing • Corticotropin • Danazol • Dextrothyroxine • Diuretics, thiazide, thiazide-related • Epinephrine • Estrogens • Ethacrynic acid • Furosemide • Glucagon • Molindone • Phenytoin • Thyroid hormones • Triamterene

Antimicrobials

Cardiovascular Agents

Antiosteo-porosis Agents

Antiplatelets

Diabetes Agents

Ulcer Treatment

Diuretics

Other Agents

Glyburide

Immuno-suppression

Antimicrobials

Cardiovascular Agents

Antiosteo-porosis Agents

Antiplatelets

Diabetes Agents

Ulcer Treatment

Diuretics

Other Agents

Brand Name	Diaβeta®	Micronase®	Glynase® PresTab®
Company	Hoechst Marion Roussel Pharmaceuticals, Inc.	Pharmacia and Upjohn, Inc.	Pharmacia and Upjohn, Inc.
Mechanism of Action	• Decreases blood glucose by stimulating the release of insulin from the pancreas in patients with functioning beta cells in the pancreatic islets		
Indication	• Adjunct to diet to lower blood glucose in patients with non-insulin-dependent diabetes mellitus (Type II) whose hyperglycemia cannot be controlled by diet alone		
Contraindications	• Known hypersensitivity or allergy to agent • Diabetic ketoacidosis, with or without coma • Type I diabetes mellitus, as sole therapy (Micronase, Glynase)		
Special Consideration	• Oral hypoglycemic drugs have been associated with increased cardiac mortality as compared to treatment with diet alone or diet and insulin		
Adverse Reactions	• Hypoglycemia • Cholestatic jaundice and hepatitis (therapy should be discontinued) • Liver function abnormalities • Gastrointestinal disturbances (nausea, epigastric fullness, heartburn) • Dermatologic reactions (pruritus, erythema, urticaria, morbilliform) • Hematologic reactions (leukopenia, agranulocytosis, thrombocytopenia, hemolytic anemia, aplastic anemia, pancytopenia) • Metabolic reactions (hepatic porphyria, disulfiram-like reactions) • Hyponatremia		
Drug Interactions	• NSAIDs • Salicylates • Sulfonamides • Chloramphenicol • Probenecid • Coumarins • MAO inhibitors • Beta-blockers • Thiazides and other diuretics • Corticosteroids • Phenothiazines • Thyroid products • Estrogens • Oral contraceptives • Phenytoin • Nicotinic acid • Sympathomimetics • Calcium channel blockers • Isoniazid • Miconazole		

Continued ...

Immuno-suppression

Antimicrobials

Cardiovascular Agents

Antiosteo-porosis Agents

Antiplatelets

Diabetes Agents

Ulcer Treatment

Diuretics

Other Agents

Glyburide

Dosage	*Diaβeta* • Usual starting dose range is 2.5 mg/d to 5 mg/d administered with breakfast or the first main meal • Usual maintenance range is 1.25 mg/d to 20 mg/d given as a single dose or in divided doses • Increments should be no greater than 2.5 mg at weekly intervals *Micronase* • Usual starting dose range is 2.5 mg/d to 5 mg/d administered with breakfast or the first main meal • Usual maintenance range is 1.25 mg/d to 20 mg/d given as a single dose or in divided doses • Increments should be no greater than 2.5 mg at weekly intervals *Glynase* • Patients should be retitrated when transferred from other oral hypoglycemic agents • Usual starting dose range is 1.5 mg/d to 3 mg/d administered with breakfast or the first main meal • Usual maintenance range is 0.75 mg/d to 12 mg/d given as a single dose or in divided doses • Increments should be no greater than 1.5 mg at weekly intervals

Editors' Notes:

Glyburide may be given after pancreas transplantation in patients with impaired glucose tolerance.

Glipizide

Immuno-suppression

Brand Name	Glucotrol®
Company	Pfizer
Mechanism of Action	• Stimulation of insulin secretion from beta cells of pancreatic islet tissue in patients with functioning beta cells
Indication	• Adjunct to diet for the control of hyperglycemia and its associated symptomatology in patients with non-insulin-dependent diabetes mellitus (Type II) whose hyperglycemia cannot be controlled by diet alone
Contraindications	• Known hypersensitivity or allergy to agent • Diabetic ketoacidosis, with or without coma
Special Precautions	• Oral hypoglycemic drugs have been associated with increased cardiac mortality as compared to treatment with diet alone or diet and insulin
Adverse Reactions	• Hypoglycemia • Gastrointestinal disturbances (nausea, diarrhea, constipation, gastralgia) • Dermatologic reactions (erythema, morbilliform or maculopapular eruptions, urticaria, pruritus, and eczema) • Hematologic reactions (leukopenia, agranulocytosis, thrombocytopenia, hemolytic anemia, aplastic anemia, pancytopenia) • Metabolic reactions (hepatic porphyria, disulfiram-like reactions) • Hyponatremia • Elevated liver enzymes
Drug Interactions	• NSAIDs • Salicylates • Sulfonamides • Chloramphenicol • Probenecid • Coumarins • MAO inhibitors • Beta-blockers • Thiazides and other diuretics • Corticosteroids • Phenothiazines • Thyroid products • Estrogens • Oral contraceptives • Phenytoin • Nicotinic acid • Sympathomimetic • Calcium channel blockers • Isoniazid • Miconazole
Dosage	• Recommended initial dose is 5 mg/d given before breakfast • Dosage adjustments should be made in increments of 2.5 mg to 5 mg • Maximum once-daily dose is 15 mg; dosages above 15 mg should be divided, given before meals of adequate caloric content, and should not exceed 40 mg/d

Antimicrobials

Cardiovascular Agents

Antiosteo-porosis Agents

Antiplatelets

Diabetes Agents

Ulcer Treatment

Diuretics

Other Agents

Metformin Hydrochloride

Immuno-suppression

Antimicrobials

Cardiovascular Agents

Antiosteo-porosis Agents

Antiplatelets

Diabetes Agents

Ulcer Treatment

Diuretics

Other Agents

Brand Name	**Glucophage®**
Company	Bristol-Myers Squibb Company
Mechanism of Action	• Bypasses pancreas and has direct antihyperglycemic action
Indication	• First-line therapy for the control of hyperglycemia and its associated symptomatology in patients with non-insulin-dependent diabetes mellitus (Type II)
Contraindications	• Renal disease or renal dysfunction • Should be withheld in patients undergoing radiologic studies involving parenteral administration of iodinated contrast materials • Known hypersensitivity to metformin hydrochloride • Acute or chronic metabolic acidosis, including diabetic ketoacidosis, with or without coma
Warnings	• Lactic acidosis is a serious metabolic complication that can occur due to metformin accumulation • Oral antidiabetic agents have been associated with an increased risk of cardiovascular mortality as compared to treatment with diet alone or diet and insulin
Special Precautions	• Renal function should be monitored to minimize risk of metformin accumulation • If hypoxic states occur, discontinue drug therapy • Therapy should be suspended for surgical procedures • Alcohol is known to potentiate the effect of metformin on lactate metabolism—patients should be warned about the dangers of excessive alcohol intake
Adverse Reactions	• Lactic acidosis • Gastrointestinal disturbances (nausea, vomiting, bloating, flatulence) • Dermatologic reactions (rash/dermatitis) • Hematologic reactions (asymptomatic subnormal serum vitamin B_{12} levels) • Special senses (slight metallic taste upon initiation of therapy)
Drug Interactions	• Furosemide • Nifedipine • Cationic drugs • Drugs that tend to produce hyperglycemia (thiazide and other diuretics, corticosteroids, phenothiazines, thyroid products, estrogens, oral contraceptives, phenytoin, nicotinic acid, sympathomimetics, calcium channel blockers, isoniazid)
Dosage	• Recommended starting dosage is 500 mg twice daily or 850 mg daily with meals • Dosage should be increased by one 500 mg tablet each week, or by one 850 mg tablet every 2 weeks • Individualize dosage based on effectiveness and tolerance, up to a maximum of 2550 mg/d given in divided doses

Editors' Notes:

The major concern is the development of lactic acidosis, which can be fatal. There is little information in the transplant literature regarding metformin.

Pioglitazone

Brand Name	Actos®
Company	Takeda Pharmaceuticals America, Inc. and Eli Lilly and Company
Class	Thiazolidinedione antidiabetic agent
Mechanism of Action	• Potent and highly selective agonist for peroxisome proliferator-activated receptor-gamma (PPAR(gamma)) which decreases insulin resistance in the periphery and in the liver resulting in increased insulin-dependent glucose disposal and decreased hepatic glucose output.
Indication	• Monotherapy and in combination with a sulfonylurea, metformin, or insulin when diet and exercise plus the single agent does not result in adequate glycemic control in patients with type 2 diabetes (non-insulin-dependent diabetes millitus, NIDDM).
Contraindications	• Patients with known hypersensitivity to this product or any of its components
Warnings	• Fluid retention which may lead to or exacerbate heart failure • Not recommended in patients with NYHA class 3 and 4 heart failure
Special Precautions	• Pregnancy Category C • Hypoglycemia in combination with insulin or oral hypoglycemic agents • Ovulation • Decreases in hemoglobulin and hematocrit • Idiosyncratic hepatoxicity • Weight gain • Edema
Adverse Reactions	• Headache • Pharyngitis • Sinusitis • Myalgia • Tooth Disorder • Possible increase in CPK levels
Drug Interactions	• Oral contraceptives • Ketoconazole • Midazalam • Possible cytochrome P450 isoform CYP3A4
Dosage	• Initial 15 mg or 30 mg once dally, maximum 45 mg once daily

Immuno-suppression

Antimicrobials

Cardiovascular Agents

Antiosteo-porosis Agents

Antiplatelets

Diabetes Agents

Ulcer Treatment

Diuretics

Other Agents

Rosiglitazone

Brand Name	Avandia®
Company	GlaxoSmithKline
Class	Thiazolidinedione antidiabetic agent
Mechanism of Action	• Potent and highly selective agonist for peroxisome proliferator-activated receptor-gamma (PPAR(gamma)) which decreases insulin resistance in the periphery and in the liver resulting in increased insulin-dependent glucose disposal and decreased hepatic glucose output.
Indication	• Monotherapy and in combination with a sulfonylurea, metformin, or insulin when diet and exercise plus the single agent does not result in adequate glycemic control in patients with type 2 diabetes (non-insulin-dependent diabetes millitus, NIDDM).
Contraindications	• Patients with known hypersensitivity or allergy to agent or any of its components.
Warnings	• Fluid retention which may lead to or exacerbate heart failure • Not recommended in patients with NYHA class 3 and 4 heart failure
Special Precautions	• Hypoglycemia in combination with insulin or oral hypoglycemic agents • Pregnancy Category C • Ovulation • Decreases in hemoglobulin and hematocrit • Idiosyncratic hepatoxicity • Edema • Plasma volume expansion and pre-load-induced cardiac
Adverse Reactions	• Headache
Drug Interactions	• None identified
Formulations	• 2 mg, 4 mg and 8 mg film-coated tablets in bottles of 30, 60, 100, 500 and SUP of 100
Dosage	• Initial dose is 4 mg administered as either a single dose once daily or in divided doses twice daily • Maximum dose should not exceed 8 mg daily, as a single dose or divided twice daily

Immuno-suppression

Antimicrobials

Cardiovascular Agents

Antiosteo-porosis Agents

Antiplatelets

Diabetes Agents

Ulcer Treatment

Diuretics

Other Agents

Chapter 7
Ulcer Prophylaxis and Treatment

Ulcer Prophylaxis and Treatment Overview

**Magnesium/Aluminum Hydroxide Suspension,
Aluminum Hydroxide Suspension,
Calcium Carbonate**

Histamine H2-Receptor Antagonists

- Famotidine
- Ranitidine
- Cimetidine

Proton Pump Inhibitors

- Omeprazole
- Lansoprazole
- Pantoprazole
- Rabeprazole
- Esomeprazole

Sucralfate

Ulcer Prophylaxis and Treatment Overview

Agent	Dosage	
Magnesium/ Aluminum Hydroxide Suspension	*Maalox®* • 1 or 2 teaspoons, 4 times a day, taken 20 minutes to 1 hour after meals and at bedtime *ALternaGEL™* • 1 to 2 teaspoons, as needed, between meals, and at bedtime • May be followed by a sip of water *Amphojel®* • 2 teaspoons, followed by a sip of water, 5 to 6 times a day between meals and bedtime	
Calcium Carbonate	*Tums E-X®* • 1 or 2 tablets as symptoms occur • Patients should take no more than 10 tablets in a 24-hour period	
Histamine H2- Receptor Antagonists	*Famotidine* Duodenal—Acute Duodenal—Maintenance Benign Gastric—Acute *Ranitidine* Duodenal—Acute Duodenal—Maintenance Benign Gastric *Cimetidine* Duodenal—Acute Duodenal—Maintenance Benign Gastric	• 40 mg once a day at bedtime for up to 6 to 8 weeks • 20 mg once a day at bedtime • 40 mg once a day at bedtime • 150 mg or 10 mL twice a day —or— 300 mg or 20 mL once a day at bedtime • 150 mg or 10 mL at bedtime • 150 mg or 10 mL twice a day • 800 mg once daily at bedtime for up to 6 to 8 weeks • 400 mg at bedtime • 800 mg once daily at bedtime for up to 6 weeks
Proton Pump Inhibitors	• *Omeprazole* • *Lansoprazole* • *Pantoprozale* • *Rabeprazole* • *Esomeprazole*	• 20 or 40 mg once daily • 15 or 30 mg once daily • 40 mg once daily, orally or intravenously • 20 mg or 40 mg once or twice daily • 20 mg or 40 mg orally or IV daily
Sucralfate	*Acute Therapy* • 1 g, 4 times a day on an empty stomach	

Immuno-suppression

Antimicrobials

Cardiovascular Agents

Antiosteo-porosis Agents

Antiplatelets

Diabetes Agents

Ulcer Treatment

Diuretics

Other Agents

Immuno-suppression

Magnesium/Aluminum Hydroxide Suspension, Aluminum Hydroxide Suspension, Calcium Carbonate

Names of Commonly Used Brands	*Magnesium/Aluminum* • Maalox® • Maalox® TC	*Aluminum* •ALternaGEL™ •Amphojel®	*Calcium* •Tums E-X® Tablets
Mechanism of Action	• React chemically to neutralize or buffer existing quantities of stomach acid, resulting in an increased pH value of stomach content • Decrease acidity within lumen of the esophagus, causing an increase in intraesophageal pH and a decrease in pepsin activity		
Indication	• Symptomatic relief of hyperacidity associated with peptic acid and other gastrointestinal conditions requiring high degree of neutralization		
Contraindications/ Medical Problems	• Alzheimer's disease • Appendicitis • Undiagnosed gastrointestinal/ rectal bleeding • Colitis • Colostomy, diverticulitis, or ileostomy • Constipation • Gastric outlet obstruction • Diarrhea • Intestinal obstruction • Hemorrhoids • Renal function impairment • Sensitivity to magnesium or aluminum	• Alzheimer's disease • Appendicitis • Undiagnosed gastrointestinal/ rectal bleeding • Gastric outlet obstruction • Diarrhea • Intestinal obstruction • Hemorrhoids • Renal function impairment • Sensitivity to aluminum	• Hypercalcemia • Appendicitis • Undiagnosed gastrointestinal/rectal bleeding • Constipation • Hypoparathyroidism • Intestinal obstruction • Hemorrhoids • Renal function impairment • Sarcoidosis • Sensitivity to calcium • Diarrhea
Adverse Reactions	• Hypermagnesemia • Neurotoxicity • Fecal impaction • Hypercalcemia • Osteomalacia and osteoporosis due to phosphate depletion • Phosphorous depletion syndrome	• Neurotoxicity • Fecal impaction • Hypercalcemia • Osteomalacia and osteoporosis due to phosphate depletion • Phosphorous depletion syndrome	• Fecal impaction • Metabolic alkalosis • Hypercalcemia • Renal calculi • Osteomalacia

Continued ...

Antimicrobials · Cardiovascular Agents · Antiosteoporosis Agents · Antiplatelets · Diabetes Agents · Ulcer Treatment · Diuretics · Other Agents

Magnesium/Aluminum Hydroxide Suspension, Aluminum Hydroxide Suspension, Calcium Carbonate

Immuno-suppression · Antimicrobials · Cardiovascular Agents · Antiosteo-porosis Agents · Antiplatelets · Diabetes Agents · Ulcer Treatment · Diuretics · Other Agents

Names of Commonly Used Brands	*Magnesium/Aluminum* • Maalox® • Maalox® TC	*Aluminum* • ALternaGEL™ • Amphojel®	*Calcium* • Tums E-X® Tablets
Drug Interactions	• Urinary acidifiers • Amphetamines or quinidine • Anticholinergics • Chenodiol • Quinolones • Citrates • Digitalis glycosides • Enteric-coated agents • Folic acid • Histamine H$_2$-Receptor Antagonists • Oral iron preparations • Oral isoniazid • Ketoconazole • Mecamylamine • Methenamine • Misoprostol • Pancrelipase • Penicillamine • Phenothiazines • Phenytoin • Oral phosphates • Salicylates • Sodium fluoride • SPSR • Sucralfate • Tetracyclines • Vitamin D	• Urinary acidifiers • Amphetamines or quinidine • Anticholinergics • Chenodiol • Quinolones • Citrates • Digitalis glycosides • Enteric-coated agents • Folic acid • Histamine H$_2$-Receptor Antagonists • Oral iron preparations • Oral isoniazid • Ketoconazole • Mecamylamine • Methenamine • Penicillamine • Phenothiazines • Phenytoin • Oral phosphates • Salicylates • Sodium fluoride • Sucralfate	• Tetracyclines • Urinary acidifiers • Amphetamines or quinidine • Anticholinergics • Calcitonin • Cellulose sodium phosphate • Quinolones • Citrates • Diuretics • Enteric-coated agents • Histamine H$_2$-Receptor Antagonists • Oral iron preparations • Ketoconazole • Mecamylamine • Methenamine • Milk products • Pancrelipase • Phenytoin • Oral phosphates • Salicylates • Sodium bicarbonate • Sodium fluoride • SPSR • Sucralfate • Tetracyclines
Dosage	*Maalox®* • 1 or 2 teaspoons, 4 times a day, taken 20 minutes to 1 hour after meals and at bedtime	*ALternaGEL™* • 1 to 2 teaspoons, as needed, between meals, and at bedtime • May be followed by a sip of water *Amphojel®* • 2 teaspoons, followed by a sip of water, 5 to 6 times a day between meals and at bedtime	*Tums E-X®* • 1 or 2 tablets as symptoms occur • Patients should take no more than 10 tablets in a 24-hour period

Histamine H2-Receptor Antagonists

Brand Name	Pepcid® (famotidine)	Zantac® (Ranitidine)	Tagamet® (cimetidine)
Company	Merck & Co., Inc.	Glaxo Wellcome, Inc.	SmithKline Beecham Pharmaceuticals
Mechanism of Action	• Inhibit basal and nocturnal gastric secretions by competitively and reversibly inhibiting the action of histamine at the histamine H_2-receptors of parietal cells		
Indications	• Short-term treatment of active duodenal ulcer • Maintenance treatment of duodenal ulcer at reduced dosage after healing of active ulcer • Short-term treatment of active benign gastric ulcer		
Contraindication	• Hypersensitivity to product		
Adverse Reactions	• Headache • Dizziness • Constipation • Diarrhea	• *CNS* Headache, and rarely, malaise, dizziness, somnolence, insomnia, vertigo, reversible blurred vision, reversible involuntary motor disturbances • *CV* Arrhythmias • *GI* Constipation, diarrhea, nausea/vomiting, abdominal pain • *Hepatic* Elevated enzymes • *Integumental* Rash	• *CNS* Headache, dizziness, somnolence, insomnia, vertigo, reversible confused states • *CV* Arrhythmias • *Endocrine* Gynecomastia • *GI* Diarrhea • *Hepatic* Elevated enzymes • *Integumental* Rash • *Renal* Elevated serum creatinine
Drug Interactions	• None identified	• None identified at recommended doses	• Warfarin • Phenytoin • Propranolol • Nifedipine • Chlordiazepoxide • Diazepam • Certain tricyclic antidepressants • Lidocaine • Theophylline • Metronidazole

Continued ...

Immuno-suppression

Antimicrobials

Cardiovascular Agents

Antiosteo-porosis Agents

Antiplatelets

Diabetes Agents

Ulcer Treatment

Diuretics

Other Agents

Histamine H2-Receptor Antagonists

Brand Name	Pepcid® (famotidine)	Zantac® (ranitidine)	Tagamet® (cimetidine)
Dosage	*Duodenal—Acute* • 40 mg once a day at bedtime for up to 6 to 8 weeks *Duodenal— Maintenance* • 20 mg once a day at bedtime *Benign Gastric—Acute* • 40 mg once a day at bedtime	*Duodenal—Acute* • 150 mg or 10 mL twice a day —or— 300 mg or 20 mL once a day at bedtime *Duodenal— Maintenance* • 150 mg or 10 mL at bedtime *Benign Gastric* • 150 mg or 10 mL twice a day	*Duodenal—Acute* • 800 mg once daily at bedtime for up to 6 to 8 weeks *Duodenal— Maintenance* • 400 mg at bedtime *Benign Gastric* • 800 mg once daily at bedtime for up to 6 weeks

Editors' Notes:

Cimetidine may increase cyclosporine levels and cause nephrotoxicity. Famotidine, ranitidine, and cimetidine are now available over the counter.

Immuno-suppression

Antimicrobials

Cardiovascular Agents

Antiosteo-porosis Agents

Antiplatelets

Diabetes Agents

Ulcer Treatment

Diuretics

Other Agents

Proton Pump Inhibitors

Immuno-suppression

Antimicrobials

Cardiovascular Agents

Antiosteo-porosis Agents

Antiplatelets

Diabetes Agents

Ulcer Treatment

Diuretics

Other Agents

Brand Name	Prilosec® (Omeprazole)
Company	Astra / Merck Group
Mechanism of Action	• Suppresses basal and stimulated gastric acid secretion by specific inhibition of the H^+/K^+ ATPase enzyme system at the secretory surface of the gastric parietal cell
Indication	• Short-term treatment of active duodenal ulcer, gastric ulcer, symptomatic GERD, erosive esophagitis • Pathological hypersecretory conditions • Treatment of H.pylori in combination with other agents
Contraindication	• Known hypersensitivity to any component of this agent
Adverse Reactions	• Headache • Diarrhea • Abdominal pain • Nausea • URI • Dizziness • Vomiting • Rash • Constipation • Cough • Asthenia • Back pain
Drug Interactions	• Diazepam • Warfarin • Phenytoin • Drugs metabolized via the cytochrome P-450 system (ie, cyclosporine, disulfiram)
Formulations	• 10 mg delayed-release capsules • 20 mg delayed-release capsules • 40 mg delayed-release capsules • 20 mg delayed-release tablets
Dosage	• 20 or 40 mg once daily

Proton Pump Inhibitors

Brand Name	® (Lansoprazole)
Company	TAP Pharmaceuticals
Mechanism of Action	• Suppresses basal and stimulated gastric acid secretion by specific inhibition of the H+/K+ ATPase enzyme system at the secretory surface of the gastric parietal cell
Indications	• Short-term treatment of active duodenal ulcer—15 mg once daily for 4 weeks • Combination therapy for *H. pylori* eradication to reduce the risk of duodenal ulcer recurrence—30 mg 2 to 3 times daily for 10 to 14 days • Maintenance of healed duodenal ulcers—15 mg once daily • Short-term treatment of active benign gastric ulcers—30 mg once daily for up to 8 weeks • Short-term treatment of symptomatic gastroesophageal reflux disease—15 mg once daily for up to 8 weeks • Short-term treatment of erosive esophagitis—30 mg once daily for up to 8 to 16 weeks • Maintenance of healing of erosive esophagitis—15 mg once daily • Pathological hypersecretory conditions including Zollinger-Ellison Syndrome—initial doses of 60 mg once daily, adjusted to patients need, up to 90 mg twice daily have been administered
Contraindication	• Hypersensitivity to any component of the formulation
Adverse Reactions	• Headache • Diarrhea
Drug Interactions	• Theophylline • Sucralfate
Formulations	• 15 mg delayed-release capsules • 30 mg delayed-release capsules • 15 mg orally disintegrated, delayed-release tablets • 30 mg orally disintegrated, delayed-release tablets • 30 mg lyophilized powder for injection
Dosage	• 15 or 30 mg once daily

Proton Pump Inhibitors

Brand Name	**Protonix® (pantoprazole)**
Company	Wyeth-Ayerst
Class	Proton pump inhibitor
Mechanism of Action	• Suppresses basal and stimulated gastric acid secretion by specific inhibition of the H^+/K^+ ATPase enzyme system at the secretory surface of the gastric parietal cell
Indications	• Short-term treatment of erosive esophagitis associated with Gastroesophageal reflux (GERD) • The oral form is indicated for the short-term treatment (up to 8 weeks) in the healing and symptomatic relief of erosive esophagitis. For those patients who have not healed after 8 weeks of treatment, and additional 8 week course may be considered. • IV form is indicated for short-term treatment (7 to 10 days) of gastroesophageal reflux disease (GERD), as an alternative to oral therapy in patients who are unable to continue taking the oral form. Safety and efficacy of injectable form as an initial treatment for GERD have not been demonstrated. *Treatment of Erosive Esophagitis* • The oral dose is 40 mg given once daily for up to 8 weeks. For those patients who may have not healed after 8 weeks of treatment, and additional 8-week course may be considered. Intravenous infusion is 40 mg given once daily for 7 to 10 days
Contraindication	• PROTONIX I.V. for Injection is contraindicated in patients with known hypersensitivity to the formulation.
Warning	• Pregnancy Category B
Adverse Reactions	• Headache • Rhinitis • Dizziness • Diarrhea • Rash • Dyspepsia • Nausea • Chest pain • Constipation • Vomiting • Arthralgia • Hyperglycemia • Hyperlipidemia • Liver function abnormalities • Injection site reaction
Drug Interactions	• May interfere with the absorption of drugs where gastric pH is an important determinant of their bioavailability
Formulations	• IV one carton containing 25, freeze dried 40 mg vials • PO 40 mg delayed-release tablets in bottles of 90
Dosage	• 40 mg once daily Patients should be cautioned that tablets should not be split, chewed or crushed. Injection admixtures (40 mg in 10 mL of normal saline) should be administered intravenously over at least 2 minutes

Additional Proton Pump Inhibitors	*Brand*	*Generic*	*Company*	*Dosage Range*
	Aciphex ®	Rabeprazole	Eisai	20 mg orally once or twice daily
	Nexium ®	Esomeprazole	AstraZeneca	20 mg or 40 mg orally or IV daily

Sucralfate

Brand Name	Carafate®
Company	Axcan Scandipharm
Mechanism of Action	• Forms an ulcer-adherent complex that covers the ulcer site and protects it from further attack by acid, pepsin, and bile salts
Indications	• Short-term treatment of active duodenal ulcers • Maintenance therapy for duodenal ulcers at reduced dosage after healing of acute ulcers
Contraindication	• None known
Adverse Reactions	• Constipation • Diarrhea • Nausea/vomiting • Gastric discomfort • Indigestion • Flatulence • Dry mouth • Pruritus • Rash • Dizziness • Sleepiness • Vertigo • Back pain • Headache
Drug Interactions	• Cimetidine • Phenytoin • Ciprofloxacin • Ranitidine • Digoxin • Tetracycline • Norfloxacin • Theophylline
Dosage	*Acute Therapy* • 1 g, 4 times a day on an empty stomach

Immuno-suppression

Antimicrobials

Cardiovascular Agents

Antiosteo-porosis Agents

Antiplatelets

Diabetes Agents

Ulcer Treatment

Diuretics

Other Agents

Chapter 8
Diuretics

Diuretics Overview

Diuretics

- Furosemide
- Bumetanide

Diuretics Overview

Agent	Dosage
Furosemide	*Oral* • Dosage should be individualized • Initial dose is 20 mg to 80 mg given as a single dose • Dose can be raised by 20 mg or 40 mg, given no sooner than 6 to 8 hours after the previous dose, until desired diuretic effect has been achieved *Parenteral* • Initial dose is 20 mg to 40 mg given as a single dose, IM or IV. IV injection should be given over 1 to 2 minutes • Dose can be raised by 20 mg, given no sooner than 2 hours after the previous dose, until desired diuretic effect has been achieved *Acute Pulmonary Edema* • Initial dose is 40 mg given intravenously over 1 to 2 minutes • Dose can be increased to 80 mg given intravenously if a satisfactory response has not occurred within 1 hour
Bumetanide	*Oral* • Dosage should be individualized • Initial dosage is 0.5 mg to 2.0 mg given as a single dose on alternate days, or for 3 to 4 days with rest periods of 1 to 2 days *Parenteral* • Initial dose is 0.5 mg to 1.0 mg, IM or IV. IV injection should be given over 1 to 2 minutes

Immuno-suppression

Antimicrobials

Cardiovascular Agents

Antiosteo-porosis Agents

Antiplatelets

Diabetes Agents

Ulcer Treatment

Diuretics

Other Agents

Diuretics

Immuno-suppression

Antimicrobials

Cardiovascular Agents

Antiosteo-porosis Agents

Antiplatelets

Diabetes Agents

Ulcer Treatment

Diuretics

Other Agents

Brand Name	Lasix® (furosemide)	Bumex® (bumetanide)
Company	Hoechst Marion Roussel Pharmaceuticals Inc.	Roche Laboratories
Mechanism of Action	• Primarily inhibits the absorption of sodium and chloride in the proximal and distal tubules, as well as in the loop of Henle	• Inhibits sodium reabsorption in the ascending limb of the loop of Henle, as shown by marked water reduction of free-water clearance during hydration and tubular free-water reabsorption during hydropenia
Indication	• Edema associated with congestive heart failure, cirrhosis of the liver, and renal disease	
Contraindications	• Anuria • Hypersensitivity to agent	• Anuria • Hepatic coma or states of severe electrolyte depletion • Hypersensitivity to agent
Adverse Reactions	• *GI* Pancreatitis, jaundice, anorexia, oral and gastric irritation, cramping, diarrhea, constipation, nausea, vomiting • *CNS* Tinnitus and hearing loss, paresthesia, vertigo, dizziness, headache, blurred vision, xanthopsia • *Hematologic* Aplastic anemia, thrombocytopenia, agranulocytosis, hemolytic anemia, leukopenia, anemia • *Dermatologic* Exfoliative dermatis, erythema multiforme, purpura, photosensitivity, urticaria, rash, pruritus • *Other* Orthostatic hypotension, hyperglycemia, glycosuria, hyperuricemia, muscle spasm, weakness, restlessness, urinary bladder spasm, thrombophlebitis, IM injection site pain, fever	• Muscle cramps • Dizziness • Hypotension • Headache • Nausea • Encephalopathy

Continued ...

Diuretics

Brand Name	Lasix® (furosemide)	Bumex® (bumetanide)
Drug Interactions	• Aminoglycosides • Ethacrynic acid • Salicylates (high doses) • Succinylcholine • Lithium • Other antihypertensive agents • Norepinephrine • NSAIDs • Indomethacin	• Aminoglycosides • Drugs with nephrotoxic potential • Lithium • Probenecid • Other antihypertensive agents • Indomethacin
Dosage	*Oral* • Dosage should be individualized • Initial dose is 20 mg to 80 mg given as a single dose • Dose can be raised by 20 mg or 40 mg, given no sooner than 6 to 8 hours after the previous dose, until desired diuretic effect has been achieved *Parenteral* • Initial dose is 20 mg to 40 mg given as a single dose, IM or IV. IV injection should be given over 1 to 2 minutes • Dose can be raised by 20 mg, given no sooner than 2 hours after the previous dose, until desired diuretic effect has been achieved *Acute Pulmonary Edema* • Initial dose is 40 mg given intravenously over 1 to 2 minutes • Dose can be increased to 80 mg given intravenously if a satisfactory response has not occurred within 1 hour	*Oral* • Dosage should be individualized • Initial dosage is 0.5 mg to 2.0 mg given as a single dose on alternate days or for 3 to 4 days with rest periods of 1 to 2 days *Parenteral* • Initial dose is 0.5 mg to 1.0 mg, IM or IV. IV injection should be given over 1 to 2 minutes

Editors' Notes:

Loop diuretics elevate uric acid. Cyclosporine also elevates uric acid, which can lead to gout.

Immuno-suppression

Antimicrobials

Cardiovascular Agents

Antiosteo-porosis Agents

Antiplatelets

Diabetes Agents

Ulcer Treatment

Diuretics

Other Agents

Chapter 9
Other Concomitant Agents

Other Concomitant Agents Overview

Other Cardiovascular Agents

- Digoxin
- Isosorbide Mononitrate
- Midodrine

Erectile Dysfunction Agents

Levothyroxine

Sodium Bicarbonate

Aquaphor® Ointment

Octreotide

Metoclopramide

Tolterodine

Bethanechol

Tamsulosin Hydrochloride

Phenytoin

Warfarin

Pentoxifylline

Fludrocortisone

Potassium

Other Concomitant Agents Overview

Immuno-suppression

Antimicrobials

Cardiovascular Agents

Antiosteo-porosis Agents

Antiplatelets

Diabetes Agents

Ulcer Treatment

Diuretics

Other Agents

Agent	Dosage
Other Cardiovascular Agents	*Digoxin* • Dosage should be individualized according to patient, dosing formulation, and disease state 　Loading 　　• Peak body digoxin stores of 8 µg/kg to 12 µg/kg should provide therapeutic effect with minimum risk of toxicity for patients with heart failure and normal sinus rhythm 　　• 10 µg/kg to 15 µg/kg may be needed in patients with atrial flutter or fibrillation 　Maintenance 　　• The following formula can be used to calculate maintenance dose $$\text{Loading dose} \times \frac{\% \text{ Daily loss}}{100}$$ % Daily loss = 14 + Ccr/5 *Isosorbide Mononitrate* • 20 mg (1 tablet), 2 times a day, given 7 hours apart *Midodrine* • 10 mg, 3 times daily
Erectile Dysfunction Agents	*Sildenafil citrate* • 25-100 mg 1 hr prior to sexual activity *Tadalafil* • 5-20 mg orally (maximum frequency is once daily) *Vardenafil* • 10 or 20 mg 1 hr prior to sexual activity (maximum frequency is once daily)
Levothyroxine	• Dosage and rate of administration must be individualized *Synthroid®* Hypothyroidism 　• Usual starting dose is 1.6 µg/kg/d, administered once daily 　• 200 µg/d is usually highest maintenance dose required Myxedema Coma 　• 300 µg to 500 µg TSH Suppression 　• Initial dose is followed by daily IV doses of 75 µg to 100 µg until the patient is stable *Levothroid®* Hypothyroidism 　• Usual starting dose is 50 µg/d, with increments of 50 µg/d every 2 to 4 weeks 　• 100 µg/d to 200 µg/d is usual maintenance range Myxedema Coma 　• 200 µg to 500 µg given rapidly

Continued ...

Other Concomitant Agents Overview

Immuno-suppression

Antimicrobials

Cardiovascular Agents

Antiosteo-porosis Agents

Antiplatelets

Diabetes Agents

Ulcer Treatment

Diuretics

Other Agents

Agent	Dosage
Levothyroxine (continued)	*Levoxine®* Hypothyroidism • Usual starting dose is 25 µg/d to 100 µg/d • A maintenance dose of 100 µg/d to 200 µg/d can be achieved within 2 to 3 weeks Myxedema Coma • Dose may be as little as 12.5 µg/d, with incremental increases of 25 µg/d at 3 to 4 week intervals
Sodium Bicarbonate	• 1 to 4 tablets QID as needed to replace bicarbonate loss
Aquaphor® Ointment	• Applied liberally to affected areas 2 to 3 times a day
Octreotide	• Initial dosage is 50 µg/d, administered subcutaneously once or twice daily • Number of injections and dosage may be increased gradually based on patient response
Metoclopramide	*Symptomatic Gastroesophageal Reflux* • 10 mg to 15 mg orally up to QID, 30 minutes before each meal and at bedtime for no longer than 12 weeks *Diabetic Gastric Stasis* • 10 mg, 30 minutes before each meal and at bedtime for 2 to 8 weeks *Nausea and Vomiting Associated With Cancer Chemotherapy* • 1 mg/kg to 2 mg/kg IV, infused slowly over a period not less than 15 minutes, 30 minutes before beginning chemotherapy and repeated every 2 hours for 2 doses and then every 3 hours for 3 doses *Postoperative Nausea and Vomiting* • 10 mg to 20 mg IM, near the end of surgery *Facilitate Small Bowel Intubation* • Single 10 mg IV dose administered slowly over a 1- to 2-minute period
Tolterodine	• 1 or 2 mg tablet orally twice daily • 2 or 4 mg extended-release capsule daily
Bethanechol	*Oral* (Urecholine®, Duvoid®) • 10 mg to 50 mg, 3 or 4 times a day *Subcutaneous (Urecholine)* • 1 mL (5 mg), although some patients respond to as little as 0.5 mL (2.5 mg) initially; minimum effective dose may be repeated 3 or 4 times a day as required
Tamsulosin Hydrochloride	• 0.4-0.8 mg orally daily

Continued ..

Other Concomitant Agents Overview

Agent	Dosage
Phenytoin	*Infatabs®* • 2 Infatabs, 3 times daily, dose then adjusted to suit individual needs *Parenteral* • Status epilepticus Loading IV dose of 10 mg/kg to 15 mg/kg administered slowly, at a rate not exceeding 50 mg/min, followed by a maintenance dose of 100 mg PO or IV every 6 to 8 hours • Neurosurgery 100 mg to 200 mg IM at approximately 4-hour intervals during surgery and continued postoperatively *Suspension* • 1 teaspoon (5 mL), 3 times daily, dose then adjusted to suit individual needs
Warfarin	• Dosage must be individualized • Usual initial dosage is 2 mg/d to 5 mg/d, with daily dosage adjustments based on the results of PT determinations • 2 mg/d to 10 mg/d is normal maintenance range
Pentoxifylline	• 1 controlled-release tablet (400 mg), 3 times a day with meals, for at least 8 weeks
Fludrocortisone	*Adrenocortical Insufficiency* • 0.1 mg/d PO *Congenital Adrenogenital Syndrome* • 0.1 mg/d PO to 0.2 mg/d PO *Idiopathic Orthostatic Hypotension* • 50 μg/d PO to 200 μg/d PO
Potassium	• Normal adult concentration of serum potassium is 3.5 mmol to 5.0 mmol or mEq per liter, with 4.5 mmol or mEq often used as a reference point • 1 g of potassium acetate provides 10.26 mEq of potassium • 1 g of potassium bicarbonate provides 10 mEq of potassium • 1 g of potassium chloride provides 13.41 mEq of potassium • 1 g of potassium citrate provides 9.26 mEq of potassium • 1 g of potassium gluconate provides 4.27 mEq of potassium

Immuno-suppression

Antimicrobials

Cardiovascular Agents

Antiosteo-porosis Agents

Antiplatelets

Diabetes Agents

Ulcer Treatment

Diuretics

Other Agents

Other Cardiovascular Agents

Immuno-suppression

Antimicrobials

Cardiovascular Agents

Antiosteo-porosis Agents

Antiplatelets

Diabetes Agents

Ulcer Treatment

Diuretics

Other Agents

Brand Name	Lanoxin® (digoxin)	Imdur® (isosorbide mononitrate)
Company	Glaxo Wellcome Inc.	Key

Mechanism of Action	• Exerts direct action on cardiac muscle and the specialized conduction system
	• Also exerts indirect actions on the autonomic nervous system, involving a vagomimetic action and baroreceptor sensitization
	• Direct and indirect actions result in:
	  • increase in force and velocity of myocardial systolic contraction
	  • slowing of heart rate
	  • decreased conduction velocity through AV node
		• Relaxes vascular smooth muscle, with consequent dilatation of peripheral arteries and veins, resulting in peripheral pooling of blood and decreased venous return to the heart
		• Ultimate effect is a reduction in left ventricular end-diastolic pressure and pulmonary capillary wedge pressure

Indications	• Heart failure • Atrial fibrillation • Atrial flutter • Paroxysmal atrial tachycardia	• Prevention of angina pectoris due to coronary artery disease
Contraindications	• Ventricular fibrillation • Hypersensitivity to agent	• Allergy to organic nitrates
Adverse Reactions	• *Cardiac* Unifocal or multiform ventricular premature contractions, ventricular tachycardia, atrioventricular dissociation, accelerated junctional rhythm, and atrial tachycardia with block • *GI* Anorexia, nausea, vomiting, diarrhea • *CNS* Visual disturbances, headache, weakness, dizziness, apathy, psychosis • *Other* Gynecomastia	• Headache • Dizziness • Nausea, vomiting

Continued ...

Other Cardiovascular Agents

Brand Name	Lanoxin® (digoxin)	Ismo® (isosorbide mononitrate)
Drug Interactions	• Potassium-depleting corticosteroids and diuretics • Calcium • Quinidine • Verapamil • Amiodarone • Propafenone • Certain antibiotics • Propantheline • Diphenoxylate • Antacids • Kaolin-pectin • Sulfasalazine • Neomycin • Cholestyramine • Certain anticancer agents • Thyroid • Sympathomimetics • Succinylcholine • Beta-blockers/calcium channel blockers	• Other vasodilators, especially alcohol • Calcium channel blockers
Dosage	• Dosage should be individualized according to patient, dosing formulation, and disease state *Loading Dose* • Peak body digoxin stores of 8 µg/kg to 12 µg/kg should provide therapeutic effect with minimum risk of toxicity for patients with heart failure and normal sinus rhythm • 10 µg/kg to 15 µg/kg may be needed in patients with atrial flutter or fibrillation *Maintenance Dose* • The following formula can be used to calculate maintenance dose $\text{Loading dose} \times \dfrac{\text{\% Daily loss}}{100}$ % Daily loss = 14 + Ccr/5	• 20 mg (1 tablet), 2 times a day, given 7 hours apart

Immuno-suppression

Antimicrobials

Cardiovascular Agents

Antiosteo-porosis Agents

Antiplatelets

Diabetes Agents

Ulcer Treatment

Diuretics

Other Agents

Other Cardiovascular Agents

Immuno-suppression

Antimicrobials

Cardiovascular Agents

Antiosteo-porosis Agents

Antiplatelets

Diabetes Agents

Ulcer Treatment

Diuretics

Other Agents

Brand Name	ProAmatine® (midodrine)
Company	Roberts Pharmaceuticals
Class	Antihypotensive
Mechanism of Action	• Forms an active metabolite, desglymidodrine, an alpha$_1$ adrenergic-agonist. Activation of the arteriolar and venous vasculature, producing and increase in vascular tone and elevation of blood pressure.
Indications	• Treatment of symptomatic orthostatic hypotension. Midodrine can cause marked elevation of supine blood. It should be used in patients whose lives are considerably impaired despite standard clinical care, including non-pharmacologic treatment (such as support stockings), fluid expansion, and lifestyle alterations.
Contraindications	• Patients with severe organic heart disease, acute renal disease, urinary retention, pheochromocytoma or thyrotoxicosis. Should not be used in patients with persistent and excessive supine hypertension.
Warning	• Can cause marked elevation of supine blood pressure (BP >200 mmHg systolic) it should be used in patients whose lives are impaired despite standard clinical care.
Precautions	• Supine and sitting hypertension • Slight slowing of the heart rate primarily due to vagal reflex. • Caution in patients with urinary retention problems, as desglymidodrine acts on the alpha-adrenergic receptors of the bladder neck • Caution in orthostatic hypotensive patients who are also diabetic, as well as those with a history of visual problems who are also taking fludrocortisone acetate, which is known to cause an increase in intraocular pressure and glaucoma • Caution in patients with renal impairment • Caution in patients with hepatic impairment • Pregnancy Category C
Adverse Reactions	• Paresthesia • Piloerection • Headache • Dysuria • Supine hypertension • Confusion • Pruritis • Chills • Nervousness • Rash • Pain • Dry mouth • Vasodilation
Drug Interactions	• Cardiac glycosides may enhance or precipitate bradycardia, A.V. block or arrhythmia • Drugs that stimulate alpha-adrenergic receptors (e.g., phenylephrine, pseudoephedrine, ephedrine, phenylpropanolamine or dihydroergotamine) • Fludrocortisone • Alpha-adrenergic blocking agents, such as prazosin, terazosin, and doxazosin, can antagonize the effects of midodrine. • Potentially, drugs that compete with active tubular secretion (e.g. metformin, cimetidine, ranitidine, procainamide, triamterene, flecainide, and quinidine).

Continued ...

Other Cardiovascular Agents

Formulations	• 2.5 mg tablets in bottles of 100 • 5 mg tablets in bottles of 100
Dosage	• 10 mg 3 times daily. Dosing should take place during the daytime hours when the patient needs to be upright, pursuing the activities of daily life. • Recommended that treatment of patients with abnormal renal function be initiated using 2.5 mg doses.

Immuno-suppression

Antimicrobials

Cardiovascular Agents

Antiosteo-porosis Agents

Antiplatelets

Diabetes Agents

Ulcer Treatment

Diuretics

Other Agents

Erectile Dysfunction Agents

Immuno-suppression

Brand Name	Viagra® (sildenafil citrate)
Company	Pfizer
Class	• Erectile dysfunction agent
Mechanism of Action	• Selectively inhibits type-V cyclic guanosine monophosphate (cGMP)-specific phosphodiesterase enzyme which increases stores of cyclic GMP and nitric oxide and causes vasodilatation to the penis, and increases blood flow into cavernosal spaces
Indications	• Treatment of erectile dysfunction
Contraindications	• Known hypersensitivity to sildenafil or any components of the tablet • Concurrent use of nitrates
Special Precautions	• Use with caution in patients with anatomical deformation of the penis or in patients who have conditions which may predispose them to priapism • Use with caution when use with alpha-blockers • Pregnancy Category B
Adverse Reactions	• Hypotension • Variceal bleeding • Myocardial infarction • Musculoskeletal pain • Rash • Headache • Flush • Dizziness • Dyspepsia • Color vision disturbances • Diarrhea • Priapism • Hemorrhoid bleeding • Nasal congestion • Esophageal ulcers • Tachyphylaxis
Drug Interactions	• Alpha-blocker • Cimetidine • Clarithromycin/erythromycin • Indinavir • Nitrate • Ketoconazole/itraconazole • Grapefruit juice
Formulations	• 25 mg tablets • 50 mg tablets • 100 mg tablets
Dosage	• 25-100 mg orally 1 hour prior to sexual activity; maximum frequency of administration is once daily

Additional Agents	Brand	Generic	Company	Dosage Range
	Cialia®	Tadalafil	Lilly	5-20 mg orally; maximum frequency of administration is once daily
	Levitra®	Vardenafil	Bayer	10 or 20 mg orally 1 hour prior to sexual activity; maximum frequency of administration once daily

Antimicrobials
Cardiovascular Agents
Antiosteo-porosis Agents
Antiplatelets
Diabetes Agents
Ulcer Treatment
Diuretics
Other Agents

Levothyroxine

Brand Name	Synthroid®	Levothroid®	Levoxine®
Company	Knoll Laboratories	Forest Laboratories, Inc.	Daniels Pharmaceuticals, Inc.
Mechanism of Action	• Provide thyroid hormone that is structurally and functionally to equivalent an endogenously produced thyroid hormone		
Indications	• Replacement or supplemental therapy for patients with hypothyroidism • Pituitary TSH suppressant • Diagnostic agent in suppression tests		• Replacement or supplemental therapy for patients with hypothyroidism
Contraindications	• Diagnosed, but uncorrected adrenal cortical insufficiency • Untreated thyrotoxicosis • Hypersensitivity to any component of drug • Myocardial infarction		
Adverse Reactions	• Adverse reactions other than those indicative of hyperthyroidism due to overdosage are rare • Signs and symptoms of hyperthyroidism include: Weight loss, palpitation, nervousness, diarrhea or abdominal cramps, sweating, tachycardia, cardiac arrhythmias, angina pectoris, tremors, headache, insomnia, intolerance to heat, and fever		
Drug Interactions	• Oral anticoagulants • Insulin or oral hypoglycemics • Cholestyramine • Estrogen, oral contraceptives		
Dosage	• Dosage and rate of administration must be individualized according to patient response *Hypothyroidism* • Usual starting dose is 1.6 µg/kg/d administered once daily • 200 µg/d is usually highest maintenance dose required *Myxedema Coma* • 300 µg to 500 µg • Initial dose is followed by daily IV doses of 75 µg to 100 µg until patient is stable *TSH Suppression* • Doses > 2 µg/kg/d are usually required	• Dosage and rate of administration must be individualized according to patient response *Hypothyroidism* • Usual starting dose is 50 µg/d, with increments of 50 µg/d every 2 to 4 weeks • 100 µg/d to 200 µg/d is usual maintenance range *Myxedema Coma* • 200 µg to 500 µg given rapidly	• Dosage and rate of administration must be individualized according to patient response *Hypothyroidism* • Usual starting dose is 25 µg/d to 100 µg/d • A maintenance dose of 100 µg/d to 200 µg/d can be achieved within 2 to 3 weeks *Myxedema Coma* • Dose may be as little as 12.5 µg/d, with incremental increases of 25 µg/d at 3 to 4 week intervals

Immuno-suppression | Antimicrobials | Cardiovascular Agents | Antiosteo-porosis Agents | Antiplatelets | Diabetes Agents | Ulcer Treatment | Diuretics | Other Agents

Immuno-
suppression

Antimicrobials

Cardiovascular
Agents

Antiosteo-
porosis Agents

Antiplatelets

Diabetes
Agents

Ulcer
Treatment

Diuretics

Other
Agents

Sodium Bicarbonate

Mechanism of Action	• Replaces bicarbonate in bladder-drained pancreas
Indication	• Relief of acid indigestion, sour stomach, or heartburn
Dosage	• 1 to 4 tablets QID as needed to replace bicarbonate loss

Editors' Notes:

After pancreas transplantation with bladder drainage, up to 10 to 12 tablets per day are necessary to prevent metabolic acidosis. Monitor CO_2 levels.

Aquaphor® Ointment

Brand Name	• Aquaphor® Ointment • Aquaphor® Antibiotic Formula • Aquaphor® Natural Healing Ointment
Company	Beiersdorf Inc.
Mechanism of Action	• Absorbs several times its own weight, forming smooth, creamy, water-in-oil emulsions
Indications	• Accelerated healing of severely dry skin, cracked skin, and minor burns • To reduce skin wound healing time and risk of infection • Follow-up skin treatment for patients undergoing radiation therapy or other drying/burning medical therapies
Precautions	• For external use only • Avoid contact with eyes • Should not be applied over third-degree burns, deep or puncture wounds, infections, or lacerations
Dosage	• Applied liberally to affected areas 2 to 3 times a day

Editors' Notes:

Very useful in diabetic patients.

Immuno-suppression

Antimicrobials

Cardiovascular Agents

Antiosteo-porosis Agents

Antiplatelets

Diabetes Agents

Ulcer Treatment

Diuretics

Other Agents

Octreotide

Immuno-suppression

Brand Name	Sandostatin®
Company	Novartis
Mechanism of Action	• Exerts pharmacological actions similar to the hormone somatostatin including: • suppression of serotonin and gastroenteropancreatic peptides (gastrin, vasoactive intestinal peptide, insulin, glucagon, secretin, motilin, and pancreatic polypeptide) • suppression of growth hormone • suppression of the LH response to GnRH
Indication	• Control of symptoms in patients with metastatic carcinoid (severe diarrhea and flushing episodes) and vasoactive intestinal peptide-secreting tumors (profuse watery diarrhea)
Contraindication	• Known hypersensitivity to any component of this agent
Adverse Reactions	• Gallstones or sludge • Nausea • Injection site pain • Diarrhea • Abdominal pain/discomfort • Loose stools • Vomiting
Drug Interactions	• Therapy used to control glycemic states (sulfonylureas, insulin, diazoxide, beta-blockers) • Agents used to control electrolyte balance • Octreotide has been associated with alterations in nutrient absorption, and thus may have an effect on any orally administered drug
Dosage	• Initial dosage is 50 µg/d, administered subcutaneously once or twice daily • Number of injections and dosage may be increased gradually based on patient response

Editors' Notes:

Prospective, randomized trial did not show a benefit after pancreas transplantation. Octreotide lowers cyclosporine-A levels and may cause escape rejection if cyclosporine levels are not carefully monitored. Intravenous infusions of 25-50 mcg/hour have been utilized.

Antimicrobials

Cardiovascular Agents

Antiosteo-porosis Agents

Antiplatelets

Diabetes Agents

Ulcer Treatment

Diuretics

Other Agents

Metoclopramide

Immuno-suppression

Antimicrobials

Cardiovascular Agents

Antiosteo-porosis Agents

Antiplatelets

Diabetes Agents

Ulcer Treatment

Diuretics

Other Agents

Brand Name	**Reglan®**
Company	Roberts Pharmaceutical Corp.
Mechanism of Action	• Increases tone and amplitude of gastric contractions, relaxes the pyloric sphincter and the duodenal bulb, and increases peristalsis of the duodenum and jejunum, possibly by sensitizing tissues to the action of acetylcholine, resulting in accelerated gastric emptying and intestinal transit
Indications	• Short-term treatment of symptomatic gastroesophageal reflux • Relief of symptoms associated with acute and recurrent diabetic gastric stasis • Prophylaxis of nausea and vomiting associated with cancer chemotherapy • Prophylaxis of postoperative nausea and vomiting • To facilitate small bowel intubation • To stimulate gastric emptying and intestinal transit of barium during radiologic examination
Contraindications	• Conditions in which stimulation of gastrointestinal motility might be dangerous (e.g., presence of gastrointestinal hemorrhage, mechanical obstruction, or perforation) • Pheochromocytoma • Epilepsy • Hypersensitivity to any component of this agent
Adverse Reactions	• *CNS* Restlessness, drowsiness, fatigue, lassitude, insomnia, headache, confusion, dizziness, mental depression • *Extrapyramidal Reactions* Acute dystonic reactions, Parkinsonian-like symptoms • *Endocrine Disturbances* Galactorrhea, amenorrhea, gynecomastia, impotence secondary to hyperprolactinemia, fluid retention secondary to transient elevation of aldosterone • *Cardiovascular* Hypotension, hypertension, supraventricular tachycardia and brady-cardia • *Gastrointestinal* Nausea and bowel disturbances, primarily diarrhea • *Hepatic* Rarely, cases of hepatotoxicity • *Renal* Urinary frequency and incontinence • *Hematologic* Few cases of neutropenia, leukopenia, or agranulocytosis • *Allergic Reactions* Few cases of rash, urticaria, or bronchospasm

Continued ...

Metoclopramide

Drug Interactions	• Anticholinergic drugs • Narcotic analgesics • Alcohol, sedatives, hypnotics, narcotics, tranquilizer • MAO inhibitors • Digoxin • Acetaminophen • Tetracycline • Levodopa • Ethanol • Insulin
Dosage	*Symptomatic Gastroesophageal Reflux* • 10 mg to 15 mg orally up to QID, 30 minutes before each meal and at bedtime for no longer than 12 weeks *Diabetic Gastric Stasis* • 10 mg, 30 minutes before each meal and at bedtime for 2 to 8 weeks *Nausea and Vomiting Associated with Cancer Chemotherapy* • 1 mg/kg to 2 mg/kg IV, infused slowly over a period not less than 15 minutes, 30 minutes before beginning chemotherapy and repeated every 2 hours for 2 doses and then every 3 hours for 3 doses *Postoperative Nausea and Vomiting* • 10 mg to 20 mg IM, near the end of surgery *Facilitate Small Bowel Intubation* • Single 10 mg IV dose administered slowly over a 1- to 2-minute period

Editors' Notes:

Extrapyramidal side effects make this drug less useful than cisapride. Neurologic toxicity is extremely common posttransplant.

Immuno-suppression

Antimicrobials

Cardiovascular Agents

Antiosteo-porosis Agents

Antiplatelets

Diabetes Agents

Ulcer Treatment

Diuretics

Other Agents

Tolterodine

Brand Name	Detrol® (Detrol LA)
Company	Pharmacia & Upjohn
Class	• Competitive muscarinic receptor antagonist
Mechanism of Action	• Prevent urinary bladder contraction by inhibiting muscarinic receptors
Indications	• Treatment of overactive bladder with symptoms of urge urinary incontinence, urgency, and frequency
Contraindications	• Hypersensitivity to tolterodine • Urinary retention, gastric retention, or uncontrolled narrow-angle glaucoma
Special Precautions	• Dosage reduction is recommended for patients with significant renal or hepatic impairment • Pregnancy Category C
Adverse Reactions	• Tachycardia • Peripheral edema • Dryness of hands and feet • Dry mouth • Constipation • Dyspepsia • Allergic/hypersensitivity reaction • Vertigo • Headache • Accommodation difficulties (blurred vision) • Hepatotoxicity (rare) • Xerphthalmia • Dysuria • Nasal congestion
Drug Interactions	• Erythromycin/clarithromycin • Cyclosporine • Itraconazole/ketoconazole • Vinblastine • Warfarin
Formulations	• 1 mg tablets • 2 mg tablets • 2 mg extended-release capsules • 4 mg extended-release capsules
Dosage	• 1 or 2 mg tablet orally twice daily, or • 2 or 4 mg extended-release capsule daily

Immuno-suppression

Antimicrobials

Cardiovascular Agents

Antiosteo-porosis Agents

Antiplatelets

Diabetes Agents

Ulcer Treatment

Diuretics

Other Agents

Immuno-suppression

Antimicrobials

Cardiovascular Agents

Antiosteo-porosis Agents

Antiplatelets

Diabetes Agents

Ulcer Treatment

Diuretics

Other Agents

Bethanechol

Brand Name	Urecholine®	Duvoid®
Company	Merck & Co., Inc.	Roberts Pharmaceutical Corp.
Mechanism of Action	• Produces the effects of a stimulated parasympathetic nervous system, increasing the tone of the detrusor urinae muscle leading to the production of a contraction strong enough to initiate micturition and empty the bladder	
Indication	• Treatment of acute postoperative and postpartum nonobstructive (functional) urinary retention and for neurogenic atony of the urinary bladder with retention	
Contraindications	• Hypersensitivity to any components of this agent • Hyperthyroidism • Peptic ulcer • Latent or active bronchial asthma • Pronounced bradycardia or hypotension • Vasomotor instability • Coronary artery disease • Epilepsy • Parkinsonism	
Adverse Reactions	• *Body as Whole* Malaise • *Digestive* Abdominal cramps or discomfort, colicky pain, nausea and belching, diarrhea, borborygmi, salivation • *Renal* Urinary urgency • *CNS* Headache • *Cardiovascular* Fall in blood pressure with reflex tachycardia, vasomotor response • *Skin* Flushing producing a feeling of warmth, sensation of heat around the face, sweating • *Respiratory* Bronchial constriction, asthmatic attacks • *Other* Lacrimation, miosis	
Drug Interaction	• Ganglion blocking compounds	
Dosage	*Oral* • 10 mg to 50 mg, 3 or 4 times a day *Subcutaneous (Urecholine)* • 1 mL (5 mg), although some patients respond to as little as 0.5 mL (2.5 mg) initially; minimum effective dose may be repeated 3 or 4 times a day as required	

Editors' Notes:

Useful for some diabetic patients with neurogenic bladder dysfunction.

Tamsulosin Hydrochloride

Brand Name	Flomax®
Company	Boehringer Ingelheim
Class	• Alpha-1 adrenergic antagonist
Mechanism of Action	• Competitively binds to the alpha-1A adrenoceptor subtype located mainly in the prostate and primarily responsible for contraction of prostate smooth muscles
Indications	• Treatment of the signs and symptoms of benign prostatic hyperplasia
Contraindications	• Known hypersensitivity to tamsulosin hydrochloride or any components of the capsule
Special Precautions	• Rule out carcinoma of the prostate prior to initiation of therapy • Pregnancy Category B
Adverse Reactions	• Rash • Allergic-type reaction • Nausea/vomiting • Diarrhea/constipation • Elevations in liver enzymes • Arthralgia/ arthritis/ myalgia • Back pain • Asthenia • Somnolence • Musculoskeletal pain • Headache • Dizziness • Amblyopia • Intraoperative floppy iris syndrome • Abnormal ejaculation • Decreased libido • Sinusitis/rhinitis • Cough • Pharyngitis
Drug Interactions	• Cimetidine • Sildenafil/tadalafil/vardenafil
Formulations	• 0.4 mg capsules
Dosage	• 0.4-0.8 mg orally once a day

Immuno-suppression

Antimicrobials

Cardiovascular Agents

Antiosteo-porosis Agents

Antiplatelets

Diabetes Agents

Ulcer Treatment

Diuretics

Other Agents

Phenytoin

Immuno-suppression

Antimicrobials

Cardiovascular Agents

Antiosteo-porosis Agents

Antiplatelets

Diabetes Agents

Ulcer Treatment

Diuretics

Other Agents

Brand Name	**Dilantin®**
Company	Parke-Davis
Mechanism of Action	• Stabilizes the threshold against hyperexcitability caused by excessive stimulation or environmental changes capable of reducing membrane sodium gradient, possibly by promoting sodium efflux from neurons
Indications	• Control of generalized tonic-clonic (grand mal) and complex partial (psychomotor, temporal lobe) seizures • Prevention of seizures during or following neurosurgery *Parenteral* • Status epilepticus of the grand mal type
Contraindication	• Hypersensitivity to phenytoin or other hydantoins
Adverse Reactions	• Nystagmus • Ataxia • Slurred speech • Decreased coordination and mental confusion • Dizziness • Insomnia • Transient nervousness • Motor twitchings • Headache
Drug Interactions	• *Increase Phenytoin Serum Levels* Acute alcohol intake, amiodarone, chloramphenicol, chlordiazepoxide, diazepam, dicumarol, disulfiram, estrogens, H_2-antagonists, halothane, isoniazid, methylphenidate, phenothiazines, phenylbutazone, salicylates, succinimides, sulfonamides, tolbutamide, trazodone • *Decrease Phenytoin Serum Levels* Carbamazepine, chronic alcohol abuse, reserpine, sucralfate, molindon hydrochloride, antacids containing calcium • *May Increase or Decrease Phenytoin Serum Levels* Phenobarbital, sodium valproate, valproic acid • *Others* Tricyclic antidepressants, corticosteroids, coumarin anticoagulants, digitoxin, doxycycline, furosemide, oral contraceptives, quinidine, rifampin, theophylline, vitamin D
Dosage	*Infatabs®* • 2 Infatabs, 3 times daily, dose then adjusted to suit individual needs *Parenteral* • Status epilepticus Loading IV dose of 10 mg/kg to 15 mg/kg administered slowly, at a rate not exceeding 50 mg/min, followed by a maintenance dose of 100 mg PO or IV every 6 to 8 hours • Neurosurgery 100 mg to 200 mg IM at approximately 4-hour intervals during surgery and continued postoperatively *Suspension* • 1 teaspoon (5 mL), 3 times daily, dose then adjusted to suit individual needs

Warfarin

Immuno-suppression
Antimicrobials
Cardiovascular Agents
Antiosteo-porosis Agents
Antiplatelets
Diabetes Agents
Ulcer Treatment
Diuretics
Other Agents

Brand Name	Coumadin®
Company	DuPont Pharma
Mechanism of Action	• Inhibits the synthesis of vitamin K-dependent coagulation factors, leading to sequential depression of Factors VII, IX, X, and II
Indications	• Prophylaxis and/or treatment of venous thrombosis and its extension, pulmonary embolism, and atrial fibrillation with embolization • Adjunct in the prophylaxis of systemic embolism after myocardial infarction
Contraindications	• Pregnancy • Hemorrhagic tendencies or blood dyscrasias • Recent or contemplated surgery • Bleeding tendencies associated with active ulceration or overt bleeding • Threatened abortion • Inadequate laboratory facilities • Spinal puncture • Major regional, lumbar block anesthesia • Malignant hypertension
Adverse Reactions	• Hemorrhage from any tissue or organ • Necrosis of skin and other tissues • Alopecia • Urticaria • Dermatitis • Fever • Nausea • Diarrhea • Abdominal cramping • Cramping • Systemic cholesterol microembolization • "Purple toes" syndrome • Cholestatic hepatic injury • Hypersensitivity reactions

Drug Interactions

Increase PT Time

• Alcohol	• Diuretics	• Narcotics, prolonged
• Allopurinol	• Disulfiram	• Pentoxifylline
• Aminosalicylic acid	• Ethacrynic acid	• Phenylbutazone
• Amiodarone HCl	• Fenoprofen	• Phenytoin
• Anabolic steroids	• Glucagon	• Propafenone
• Anesthetics, inhal.	• Hepatotoxic drugs	• Pyrazolones
• Antibiotics	• Ibuprofen	• Quinidine
• Bromelains	• Indomethacin	• Quinine
• Chenodiol	• Influenza vaccine	• Quinolones
• Chloral hydrate	• Lovastatin	• Ranitidine
• Chlorpropamide	• Mefenamic acid	• Salicylates
• Chymotrypsin	• Methyldopa	• Sulfinpyrazone
• Cimetidine	• Methylphenidate	• Sulfonamides
• Clofibrate	• Metronidazole	• Sulindac
• Dextran	• Miconazole	• Tamoxifen
• Dextrothyroxine	• MAO inhibitors	• Thyroid drugs
• Diazoxide	• Nalidixic acid	• Tolbutamide
• Diflunisal	• Naproxen	• TMP/SMX

Continued …

Immuno-suppression

Antimicrobials

Cardiovascular Agents

Antiosteo-porosis Agents

Antiplatelets

Diabetes Agents

Ulcer Treatment

Diuretics

Other Agents

Warfarin

Drug Interactions (Continued)	*Decrease PT Time*
	• Adrenocortical steroids • Glutethimide
	• Alcohol • Griseofluvin
	• Aminoglutethimide • Haloperidol
	• Antacids • Meprobamate
	• Antihistamines • Nafcillin
	• Barbiturates • Oral contraceptives
	• Carbamazepine • Paraldehyde
	• Chloral hydrate • Primidone
	• Chlordiazepoxide • Ranitidine
	• Cholestyramine • Rifampin
	• Diuretics • Sucralfate
	• Diets high in vitamin K • Trazodone
	• Ethchlorvynol • Vitamin C
Dosage	• Dosage must be individualized
	• Usual initial dosage is 2 mg/d to 5 mg/d, with daily dosage adjustments based on the results of PT determinations
	• 2 mg/d to 10 mg/d is normal maintenance range

Pentoxifylline

Brand Name	Trental®
Company	Hoechst Marion Roussel Pharmaceuticals Inc.
Mechanism of Action	• Improves flow properties of blood by decreasing its viscosity
Indication	• Treatment of patients with intermittent claudication on the basis of chronic occlusive arterial disease
Contraindications	• Known hypersensitivity to any component of this agent or methylxanthines (caffeine, theophylline, theobromine)
Adverse Reactions	• *Cardiovascular* Angina/chest pain, arrhythmias/palpitation, flushing • *Digestive* Abdominal discomfort, belching/flatus/bloating, diarrhea, dyspepsia, nausea, vomiting • *CNS* Agitation/nervousness, dizziness, drowsiness, headache, insomnia, tremor, blurred vision
Drug Interactions	• Anticoagulants or platelet aggregation inhibitors • Antihypertensives (because small decreases in blood pressure may occur during treatment with pentoxifylline)
Dosage	• 1 controlled-release tablet (400 mg), 3 times a day with meals, for at least 8 weeks

Immuno-suppression

Antimicrobials

Cardiovascular Agents

Antiosteo-porosis Agents

Antiplatelets

Diabetes Agents

Ulcer Treatment

Diuretics

Other Agents

Fludrocortisone

Brand Name	**Florinef®**
Company	Apothecon
Mechanism of Action	• Acts on the distal tubules to increase potassium excretion, hydrogen ion excretion, and sodium reabsorption and subsequent water retention
Indications	• Treatment of adrenocortical insufficiency, chronic primary and chronic secondary • Treatment of congenital adrenogenital syndrome • Treatment of idiopathic orthostatic hypotension • Treatment and diagnosis of acidosis in renal tubular disorders
Contraindications/ Medical Problems	• Known hypersensitivity to any component of this agent • Cardiac disease • Congestive heart failure • Peripheral edema • Renal function impairment, except in treatment of renal tubular acidosis • Acute glomerulonephritis • Hepatic function impairment • Hypothyroidism or hyperthyroidism • Chronic nephritis • Osteoporosis
Adverse Reactions	• Generalized anaphylaxis • Congestive heart failure • Dizziness • Headache • Severe or continuing hypokalemic syndrome
Drug Interactions	• Digitalis glycosides • Hepatic enzyme inducers • Hypokalemia-causing medications • Lithium • Sodium-containing medications or foods
Dosage	*Adrenocortical Insufficiency* • 0.1 mg/d PO *Congenital Adrenogenital Syndrome* • 0.1 mg/d PO to 0.2 mg/d PO *Idiopathic Orthostatic Hypotension* • 50 μg/d PO to 200 μg/d PO

Editors' Notes:

Useful for treating hyperkalemia due to tacrolimus and cyclosporine. Also useful for orthostatic hypotension in diabetic autonomic neuropathy.

Potassium

Brand Name	Various
Mechanism of Action	• Potassium is the predominant cation within cells and, as such, high intracellular potassium concentrations are necessary for numerous cellular metabolic processes
Indication	• Treatment and prophylaxis of hypokalemia
Contraindications	• Known hypersensitivity to potassium • Hyperkalemia
Adverse Reactions	• Hyperkalemia • Contact irritation of the alimentary tract • Diarrhea • Nausea • Stomach pain, discomfort, or gas • Vomiting
Drug Interactions	• Amphotericin B • Corticosteroids • Corticotropin • Gentamicin • Penicillins • Polymyxin B • ACE inhibitors • NSAIDs • Beta-blockers • Blood from blood bank • Cyclosporine • Potassium-sparing diuretics • Heparin • Low-salt milk • Potassium-containing medications • Salt substitutes • Anticholinergics • Calcium salts • Digitalis glycosides in presence of heart block • Thiazide diuretics • Exchange resins, sodium cycle • Insulin • Sodium bicarbonate • Laxatives
Dosage	• Normal adult concentration of serum potassium is 3.5 mmol to 5.0 mmol or mEq per liter, with 4.5 mmol or mEq often used as a reference point • 1 g of potassium acetate provides 10.26 mEq of potassium • 1 g of potassium bicarbonate provides 10 mEq of potassium • 1 g of potassium chloride provides 13.41 mEq of potassium • 1 g of potassium citrate provides 9.26 mEq of potassium • 1 g of potassium gluconate provides 4.27 mEq of potassium

Immuno-suppression · Antimicrobials · Cardiovascular Agents · Antiosteoporosis Agents · Antiplatelets · Diabetes Agents · Ulcer Treatment · Diuretics · Other Agents